PERSONALITY, VALUES, CULTURE

Humans are complex social beings. To understand human behavior, an integrated perspective is required – one that considers both what we regularly do (our personality traits) and what motivates us (our values).

Personality, Values, Culture uses an evolutionary perspective to look at the similarities and differences in personality and values across modern societies. Integrating research on personality and human values into a functional framework that highlights their underlying compatibilities (driven by shared genetic and brain mechanisms), Fischer describes how personality is shaped by the complex interplay between genes and the environment, both over the course of human evolution and within the life span of individuals. He proposes a gene-culture coevolution model of personality and values to explain how and why people differ around the world and how genes, economics, social conditions, and climate jointly shape personality.

Ronald Fischer is a Reader at Victoria University of Wellington, New Zealand. He is Associate Editor of *Applied Psychology: An International Review* and serves on the editorial boards of a number of psychology and business journals. He has published more than 100 articles in psychology and related disciplines. He coauthored the second edition of *Understanding Social Psychology across Cultures* (2013) and has received a number of awards and prizes, including the Otto Klineberg Intercultural and International Relations Award, the Gordon Allport Intergroup Relations Award, and the Best Paper with International Application Award from the Academy of Management.

CULTURE AND PSYCHOLOGY

Series Editor
David Matsumoto, *San Francisco State University*

As an increasing number of social scientists come to recognize the pervasive influence of culture on individual human behavior, it has become imperative for culture to be included as an important variable in all aspects of psychological research, theory, and practice. Culture and Psychology is an evolving series of works that brings the study of culture and psychology into a single, unified concept.

Ute Schönpflug, *Cultural Transmission: Psychological, Developmental, Social, and Methodological Aspects*

Evert Van de Vliert, *Climate, Affluence, and Culture*

David Matsumoto and Fons J. R. van de Vijver, *Cross-Cultural Research Methods in Psychology*

Angela K.-y. Leung, Chi-yue Chiu and Ying-yi Hong, *Cultural Processes: A Social Psychological Perspective*

Romin W. Tafarodi, *Subjectivity in the Twenty-First Century: Psychological, Sociological, and Political Perspectives*

Yair Neuman, *Introduction to Computational Cultural Psychology*

John W. Berry, *Mutual Intercultural Relations*

George Ellis and Mark Solms, *Beyond Evolutionary Psychology*

Personality, Values, Culture

An Evolutionary Approach

Ronald Fischer
Victoria University of Wellington

CAMBRIDGE
UNIVERSITY PRESS

CAMBRIDGE
UNIVERSITY PRESS

University Printing House, Cambridge CB2 8BS, United Kingdom

One Liberty Plaza, 20th Floor, New York, NY 10006, USA

477 Williamstown Road, Port Melbourne, VIC 3207, Australia

314–321, 3rd Floor, Plot 3, Splendor Forum, Jasola District Centre, New Delhi – 110025, India

79 Anson Road, #06–04/06, Singapore 079906

Cambridge University Press is part of the University of Cambridge.

It furthers the University's mission by disseminating knowledge in the pursuit of education, learning, and research at the highest international levels of excellence.

www.cambridge.org
Information on this title: www.cambridge.org/9781107087156
DOI: 10.1017/9781316091944

First published 2018

Printed in the United Kingdom by Clays, St Ives plc

A catalogue record for this publication is available from the British Library.

Library of Congress Cataloging-in-Publication Data
Names: Fischer, Ronald, 1976– author.
Title: Personality, values, culture : an evolutionary approach / Ronald Fischer, Victoria University of Wellington.
Description: Cambridge, United Kingdom ; New York, NY : Cambridge University Press, 2018. | Series: Culture and psychology | Includes bibliographical references and index.
Identifiers: LCCN 2017035422 | ISBN 9781107087156 (alk. paper)
Subjects: LCSH: Ethnopsychology. | Personality and culture. | Values. | Evolutionary psychology.
Classification: LCC GN504 .F57 2018 | DDC 155.8/2 – dc23
LC record available at https://lccn.loc.gov/2017035422

ISBN 978-1-107-08715-6 Hardback

CONTENTS

ACKNOWLEDGMENTS

This book started as an idea generated by a presentation I gave at the 2012 Cross-Cultural Conference in Stellenbosch in South Africa. I would like to thank the series editor, David Matsumoto, for seeing the potential of this idea and then trusting and supporting me along the way to develop it further. This book represents thinking in process, and many friends and colleagues have been instrumental over the years to help me get to this point in my intellectual journey. Most importantly, Ype Poortinga has been a supportive and critical scientific mentor throughout my writing and thinking process on these topics over the last few years. Using his sharp eye and mind, he has spotted many inaccuracies and errors, and I am greatly indebted to him. I would like to thank my former students and now great friends and collaborators, especially Diana Boer and Melanie Vauclair, whose fantastic work greatly influenced my thinking. The Aarhus Institute of Advanced Studies (AIAS) and the Marie Curie-Cofund Fellowship provided generous funding, a creative space, and a stimulating intellectual atmosphere that made it possible for me to read, think, and playfully explore topics covered in this book without the pressures to provide tangible results quickly that characterize much of academia these days. My exchanges with my colleagues at AIAS were always stimulating, encouraging, and playful, whether they were conversations about quantum theory, whale communications, or Greek mythology. I especially would like to thank Djuke Veldhuis for being the perfect sparring partner for bouncing around ideas. Back home, I would like to thank my colleagues and students at Victoria University of Wellington and the Centre for Applied Cross-Cultural Research, especially Joseph Bulbulia, Colleen Ward, Garth Fletcher, and Taciano Milfont. I am grateful to my international colleagues – in particular Shalom Schwartz, Peter

Smith, Evert van de Vliert, Greg Maio, Sonia Roccas, Anat Bardi, Eldad Davidov, Jan Cieciuch, Jeff McCrae, Valdiney Gouveia, and Klaus Boehnke – for many stimulating discussions, feedback, and critical comments over the years. Velichko Fetvadjiev, Henrik Dobewall, Carolin Rabe, and Johannes Karl read various versions of the whole book and provided invaluable feedback and suggestions that I hope sharpened my arguments and writing. I am grateful to Machteld Verzijden and especially Gayathri Mohanakrishnan for countless stimulating conversations over coffee, for providing feedback on specific sections and chapters, and for opening the fascinating world of biology, evolution, and genetics for me. Many special thanks to Abigail Walkington and Janka Romero at Cambridge University Press for their help and editorial support. Finally, this book would not be in your hands without my parents and my friends, especially Liviu, Snehal, and Dilhani, who were incredibly supportive, exceptionally patient, and just simply crazy enough to keep up with my writing antics.

1

Introduction: The Quest to Understand the Person

Tina's hand is twitching furiously. Small pearls of sweat are forming on her forehead under the hot morning sun, providing a glistening contrast to her dark hair carefully tight up in a knot. Her dark eyes are wide open, her gaze fixated in trance on a distant point in the colorful crowd. Her nostrils are flaring wildly; was there not so much noise in the temple grounds, we could hear her heavy breathing. With a determined push, the priest breaks her skin and pushes the metal rod through her right cheek until it comes out through her open mouth; the only response from Tina is a hardly noticeable blink of her eyes. With a short prayer, the official inserts the next metal rod into Tina's left cheek, piercing her skin right next to one of the tiny scars from previous years and pushes it until this rod also emerges through her open mouth. Her family and friends are gathered around her, with her husband holding up a black prayer flag over her head, obscuring her from the multitude of curious onlookers. Only the shining ends of the rods protruding from her open mouth are now visible. The priest is attaching some carefully folded money and a bright-red ceremonial fruit to the end of each of the rods still innocently glistening in the early morning sun. The procedure only takes a few minutes, and Tina is now ready to perform her duties. With her hand firmly clenching the snake whip, she rises up determinedly from the plastic chair to face the crowd. In the following hours, she will wander around town, barefoot on burning-hot asphalt, blessing countless spectators and uttering short prayers in people's houses and shops. Only after having finished the prescribed route will she return with her small support group consisting of family and a few friends to the shrine to get the piercings removed. She has a long day ahead of her.

The very same day, half a world away, Sandra is also getting ready for her workday. She looks over her desk as the winter sun is slowly pushing through the clouds. The friendly faces of her parents and her brother are smiling from the framed pictures placed right next to the binders that hold all those papers related to the case she has been working hard on over the last year. Sandra pauses for a moment and smiles back at them. She has a brief flashback to last weekend, when they met at her house for dinner. She is so proud of her younger brother and his little daughter. Sandra quickly reminds herself to ask him for a photo of her niece she can put on her desk. Her eyes wander further along her desk. Piles of notes are neatly folded on the left-hand side. A tiny speck of dust is coming to rest on the black leather folder holding her application. She takes out a tissue from her handbag and wipes it off, using the moment to also quickly wipe down the rest of the neatly organized space in front of her. Her mind wanders. Five more minutes and she will face the committee. As usual on days like this, she has prepared her favorite tea and put it into the special cup that her mom gave her when she left home to go to college. After taking a final sip, she goes to the office kitchen, carefully washes the cup with soap, rinses off the soap with plenty of water, and then takes her personal towel from the drawer to wipe off the water. She hurries back to her desk and places the cup back into her locked drawer. It is time to go; it will take her three minutes to get to the meeting hall, and she wants to be there before the other committee members arrive.

Two individuals, separated by continents, one walking the streets of southern Asia, the other stepping into a meeting room in Europe. For us, Tina's behaviors may appear strange and exotic. Confronted with such behaviors, we might be shocked, revolted, mesmerized, fascinated. We may wonder what Tina is like as a person and whether she holds the same values as does Sandra. Do individuals like Tina value the same things as we do, are they motivated by the same things that makes us tick, do they think and feel about all sorts of daily activities in the same way we do? The "we" here includes you – the reader – and me, the author and observer of such fascinating rituals. Tina's behaviors seem different from ours, the cultural practices alien and hard to understand, making it more than plausible to assume that people like her have very different values and personalities from those of Sandra and our friends and colleagues back home.

In these descriptions, it is easy to focus on the specific behaviors that are richly colored through local customs and traditions. You may wonder about the meaning of Tina's holding a braided leather whip with an elaborately carved snake head. Or what Sandra's memories attached to the teacup are. My quest is to move to a different level of analysis, zooming out from specifics to detect some general patterns of human personality, searching for more or less stable predispositions that describe people across situations and for psychological processes that form the core of who we are as individuals. On this journey we may find that Tina and Sandra are actually more similar than not, even though their behavior at first glance seems so radically different.

In this book, I focus on personality traits and values. There are a number of different psychological processes I could focus on when trying to describe individuals, so I had to make choices. I define these terms more properly later, but right now it suffices to say that personality traits refer to habitual behavioral tendencies that people manifest in different situations, whereas values are goals that people find desirable and use as guides for their behavior across different situations. I use psychological research on traits and values as snapshots to glean a picture of the underlying structure of human personality. To visualize this process, let's use cars as a metaphor. Cars vary widely in colors, design, and build: some go fast; others are slow. But rather than focusing on those external features, I want to explore the differences in how they move around. In other words, what is happening under the hood? To do that we raise the hood and examine the engines that make different cars move.

Peeking under the hood of a person and examining the engines of personality, we may also wonder why people in one place of the world hold different values and show different personality traits than people in another part of the world. What motivates both Tina and Sandra to behave the way they do? You may even wonder why humans have personality at all. Why do we show such differences in our personality? Why do we value different things? Would it not make sense for us to all want the same things and behave in the same way?

The book is concerned with these and other questions about personality, values, and culture. In tackling these questions, I connect and bring together different literatures that have been used to describe individuals. There are equally rich research traditions that focus on personality traits and on values. If we want to understand what people are like and why they are behaving and acting the way they do, we need to look at both traits and values. My first question is how similar or different values and personality traits are from each other. As will become clear, even though they are studied by different research traditions that have thrived independently, traits and values are actually quite similar in many aspects. Therefore, my challenge is to bring them back together conceptually and discuss how they are related to each other.

To link values and personality traits, I must answer a second question. What are the underlying mechanisms that give rise to both values and personality traits in individuals and populations? I draw on a diverse literature in personality and social psychology research, evolutionary biology, genomics, cognitive science, and neuroscience to explore these mechanisms. Bringing these literatures together, I propose a model that can account for the similarity of values and personality traits as commonly studied in psychology. Many interesting insights have been found in recent studies investigating the role of genes, so I need to cover a lot of genetics along the way. Yet the story is even more complicated, which raises my third question. Are your values and personality traits primarily your own, or is part of who you are shaped by the complex social and cultural environment in which you live? We humans live in complex social and cultural worlds in very diverse climatic habitats around the world, which provide us not only with lots of options, opportunities, and resources but also many challenges, threats, and dangers. How does this rich tapestry influence brain processes, and how do genes regulate human psychological functions?

STUDYING PERSONALITY AROUND THE WORLD

I take a scientific approach to describing personality. My quest is to understand persons and what they are like in their environment. To that end I have gathered data from studies conducted in different parts of the world to tackle the questions of what people are like and what makes them similar and different. What kind of data are available for the science of the person? There are two methods that have been used frequently. First, researchers use systematic observations. This method is followed traditionally in the field of anthropology. Western-trained individuals observe locals in their daily behavior and report what they find interesting, striking, or theoretically relevant. However, as it may be immediately apparent, there are multiple problems with this approach; for example, can we observe all relevant behavior across multiple contexts, how do we interpret behavior like I described at the beginning of the chapter if it lies outside our personal experience, what do we pay attention to, and what do we consider worth recording? By basing our research only on our observations, we miss the personal meaning and interpretation that the protagonists give to their actions. Yet, these observations can be useful for contextualizing behavior, and so I draw on some of these studies later in this book.

A second approach is to ask people themselves what they are like as a person: this is the most popular approach taken by psychologists (Allik &

McCrae, 2004; Hofstede & McCrae, 2004; McCrae, Terracciano, & Personality Profiles of Cultures Project, 2005; Schwartz et al., 2012). The most common (because the most economical) way to ask these questions is to use standardized questionnaires. Study participants are asked to describe themselves in terms of short statements, called items, that researchers have devised based on previous research. The relevance or appropriateness of each statement to the individual is typically indicated by using a numeric scale (often called a Likert scale). For example, I will rate the item, "I am an outgoing person," in terms of the extent it describes me well (7), somewhat (4), or not at all (1). I could also chose intermediate values (2, 3, 5, 6), depending on how well I feel the statement describes me.

Of course, as do all research methods, the use of questionnaires has problems, and those differ from the problems with the observational method used in anthropology. First, the statements need to make sense, be representative of typical behavior, and be of relevance to the individual and the context (the representativeness and relevance criteria in the cross-cultural bias literature that I describe in Chapter 7). For the individual to be able to answer these questions, he or she must proceed through multiple steps. First, in any specific situation, the person has to be aware of what he or she is doing. Are Tina and Sandra aware of what they are doing at the particular moment when we caught up with them? The relevant behavior, thoughts, motivations, or beliefs need to then be memorized (encoded in memory). Across various situations and occasions, various behaviors need to be encoded and integrated into some general category that helps the individual make sense of this general class of behaviors and associated thoughts and beliefs (a process called categorization and abstraction). Both Tina and Sandra need to be able to abstract and generalize their personality observations and reflections of important values and behavior across situations and form an impression of what they are like as individuals. What values are important for Tina? What traits does Sandra use to describe her actions on a day-to-day basis? Later on, when confronted with the questionnaire by the researcher, a person has to perceive a fit between the question and some of these memorized events or abstract categories about the self (what are my important values, what are my characteristic traits?). This requires recognizing the relevance of these memories and self-categories, remembering relevant information, and choosing the subjectively appropriate answer on a scale from 1 to 5 or 1 to 7. And of course, a person needs to be motivated to engage with the questions posed by the researcher. Have a look at Figure 1.1 for a schematic description of the multiple steps involved in the overall process. In addition, much of personal description is driven by

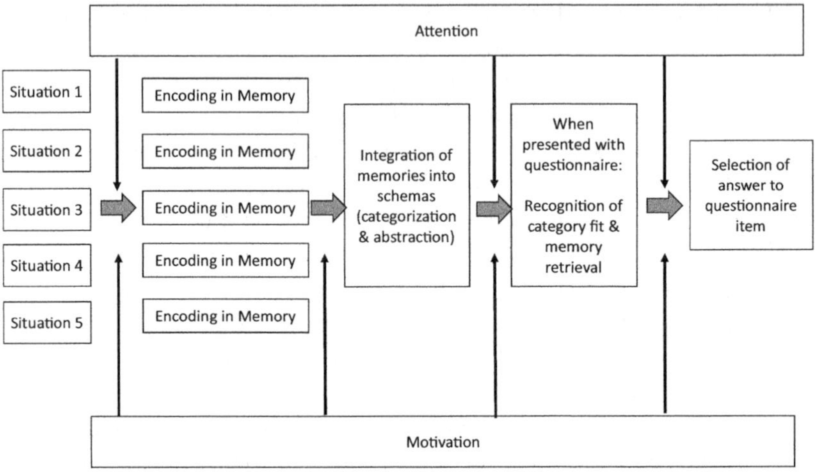

FIGURE 1.1 A schematic model showing important steps necessary to respond to personality and value questionnaires.

culturally idiosyncratic terms and phrases. I touch repeatedly on these tricky issues of language; attention to and encoding of relevant situations, behaviors, thoughts, and beliefs; and tendencies to recall them when asked to respond to survey items. When discussing personality traits and values, it is important to keep in mind how the data are gathered and the complex process that leads to these statistics that I describe and analyze throughout the book.

AN EAGLE'S-EYE VIEW ACROSS MULTIPLE LITERATURES AND ACROSS CULTURES

To bring values and personality traits together, we need to browse widely in diverse research literatures. Personality researchers have described individuals in terms of their basic behavioral dispositions, drawing not only on neuroscience, genetics, and comparative biology but also on social-cognitive theories. More recently, researchers have started to develop computational network models. Social psychologists have focused on values, building on work in sociology and anthropology. Each of these literatures is rich, multifaceted, and contradictory. Yet, all of them deal with the individual and how people think and feel about themselves. Only recently have researchers made concerted efforts to bring these diverse literatures together in the study of personality. The research on values still stands largely distinct from that body of research, and this book's aim is to integrate values into these emerging models of human personality. In the first half of the book, I present the evidence for the similarity between values and traits and provide a theoretical integration. Specifically, I bring together two different theoretical approaches: research in the tradition of the Big Five personality traits and social psychological research on values that builds on work by Milton Rokeach and Shalom Schwartz.

Turning to cultural differences, there is a large literature on culture and personality, and an even richer literature on culture and values. A casual scan of the literature suggests a relatively coherent picture. The broad consensus is that the same personality and value models apply globally, with different traits or values being more or less important in some cultures compared to others. Yet, there are empirical irregularities, niggles around the edges, and misfits, which are often glanced at and then ignored for the sake of simplicity. Hence, we need a broad view – like a bird hovering over a diverse literature. But the eyes need to be those of an eagle, eager to spot those inconsistencies and contradictions that do matter for building an integrative model of the human person. In this book, I argue that cultural differences are actually indicative of (a) a larger problem in the field's search for a simplistic model that fits all, (b) but attending to these inconsistencies is of great importance to understanding people in a global perspective, and therefore (c) attending to some of these problems is likely to yield theoretically rich new insights into the psychology of human personality. In bringing these different approaches together and drawing attention to some of the empirical inconsistencies across cultures, we will encounter a few apparent paradoxes and seemingly contradictory findings. I try to resolve these in later chapters, so I need to ask for some patience as we move through the book.

Theoretical Lumpers versus Splitters

Scientific endeavors have two broad tendencies. One is to make finer and finer distinctions between specific phenomena. The other one is to zoom out and look at a broader range of phenomena, focusing on similarities across them and general organizing principles. Some people use terms like "splitters" versus "lumpers" to describe these two broad tendencies. My approach in this book is clearly aligned with the latter. My focus is on providing a broad picture that identifies communalities that can help us make sense of human individuality across different contexts. Given the current tendencies in psychology to make finer and finer distinctions in personality and the trends in social psychology to describe smaller and smaller aspects of human psychology, we need a broad framework to examine how these various characteristics may relate to each other. The overarching framework that helps me bring these fields together is an evolutionary perspective on culture and human behavior.

When they hear the word "evolution," most people think of Darwin and of biological evolution powered by natural selection. This theory is well established, and its explanation of how different species came into existence and continue to change is well accepted. There is little controversy surrounding this explanation. Many people also accept that different individuals carry different sets of genes that make them unique. Yet, the application of evolutionary principles to explaining differences between groups of people might make some people uncomfortable. After all, history reminds us of the dangers of using scientifically flawed ideas of evolution to advance political ends to exclude and even exterminate supposedly inferior groups. It is important to use sound science and highlight ethical and practical implications.

EVOLUTION AS A GENERAL FRAMEWORK

The term "evolution" comes from the Latin *evolvere* and can be translated as "unrolling." Its broad meaning indicates a gradual development of something, typically denoting a change from a simpler form to a more complex form. The general use of the term "evolution" as indicating an unfolding or gradual development might be more acceptable to many individuals; after all, change is all around us, and even we change over the course of our individual lives. In the context of this book, I use the term "evolution" in this broad sense. Any gradual change can be driven by different processes and mechanisms. Darwin's (1859) theory focused on biological evolution, which is a specific and precise application and extension of the term to describe the change in biological species. Social systems also evolve over time, and these changes affect individuals. As with the evolution of biological species, it is possible to study changes in social systems scientifically. These evolutionary changes at the level of social systems are likely to be driven by a diverse set of variables and processes. This so-called cultural evolution shows some similarities to biological evolution, but also differs in crucial details.

Evolutionary approaches have been quite successful in providing some insights into patterns of behaviors, in particular mate selection, aggression, and cooperation. What is important here is that these approaches have focused on specific and relatively isolated behaviors. For example, researchers in economics and biology often study specific cooperation behaviors using simple games in which people have to interact with random strangers in the laboratory and need to make some financial decisions.

Or, researchers investigate changes in cultural objects such as spear tips or vases – behavioral expressions manifested in particular artifacts that can be examined in the archaeological record. It is relatively easy to study processes that lead to changes in specific behaviors such as decisions to cooperate or not or to make spear tips one way or another. There is much less work on understanding the psychological processes that might give rise to a more complex arrangement of traits and values. Some evolutionary psychologists have even dismissed attempts to study these phenomena of social change as inconsequential from a (biological) evolution perspective (Tooby & Cosmides, 1995). I disagree. As you will see in the following chapters, there are systematic patterns in the organization and variation of personality traits and values across individuals and communities. If these patterns are not random, they should be amenable to scientific analysis and explanation. A broad evolutionary framework is arguably the best-suited scientific tool for such purposes.

A BRIEF OVERVIEW OF THE BOOK

In this book, I deal with complex processes by linking various literatures to present a new approach to personality and value differences within and across cultures. Many of the studies use complex statistics and advanced mathematical models, and I distill and present their main ideas without complex equations or models. At key points, I highlight how the traits and values of our main protagonists at the beginning of this chapter can be understood from different theoretical perspectives. To prepare us for the road ahead, in Chapter 2, I outline some basic principles of evolution and how biological and cultural evolution are both similar and different. I briefly describe key terms and processes from evolutionary biology and psychology, behavioral ecology, and genetics. Researchers interested in personality traits and values are often trained in social sciences and might not be very familiar with some of these ideas coming from biology. Importantly, I introduce recent ideas of how biology and culture interact with each other through so-called gene-culture coevolution. In Chapter 3, I introduce the main ingredients of this book: personality traits and values as currently studied in psychology. Personality is a broad concept that includes many facets and aspects of what a person is like. It encompasses relative stable differences between people in their emotions, beliefs, values, goals; some researchers even include mental abilities as part of personality. It is challenging to examine all these areas across cultures simultaneously because many of these aspects have been studied infrequently and in isolation from each other, while using very diverse methods that make integration difficult. Values and personality traits (in the sense of more narrowly defined dispositional consistencies in behaviors across situations) have been researched largely independently, but there are interesting and important synergies that allow me to integrate them. I am confident that other researchers will tackle other aspects of personality and start developing a more holistic global science of personality in due course. Narrowing the scope, in Chapter 3, I present a brief history of two prototypical approaches to personality trait and value research in Western psychology. I discuss the Big Five personality trait model and the Schwartz value theory, which have been the dominant approaches in personality and social psychology, respectively. This is the core material that I use for describing personality in this book.

In Chapter 4, I examine similarities between the Big Five personality trait model and the Schwartz value theory. Values and traits are strongly influenced by the same historical roots, but have thrived in very different research traditions and literatures, with little overlap or joint discussion. My main argument in Chapter 4 is that values and traits are much more similar than many people believe. Both are organized in a similar two-dimensional structure. The organization of traits in such a two-dimensional structure might be surprising to some researchers, because interrelations between traits have been obscured by the common use of factor analysis (a statistical technique to reduce complex sets of relationships to a small number of underlying factors, which are assumed to be independent).

What might account for this great similarity of traits and values? Explaining the underlying psychological mechanics will take a few intermediate steps. In Chapter 5, I take an indirect route by examining the relative importance of both genes and culture for traits and values. A number of theoretical models have predicted that traits are more strongly influenced by genes, whereas values are more strongly shaped by culture. As I review in Chapter 5, the evidence instead suggests that both traits and values are strongly influenced by genes, whereas culture plays a much smaller role for both traits and values. In other words, overall we are more alike than different. This finding might seem surprising and unexpected. In Chapter 6, I therefore review several theoretical accounts that try to explain how personality traits (in particular) arise, drawing on recent advances in biological and neuroscience models to explain the common organization of values and personality traits. I use these insights to present a new framework that can explain the two-dimensional structure that I presented in Chapter 4.

In Chapters 4–6, I focus on genetic and biologically driven predispositional approaches. But the local environment with all its social, economic, ecological, and cultural influences needs to be brought into the picture. In Chapter 7, I begin this path by challenging the very conclusions that I reached at the end of Chapter 6. I review the evidence that shows that once we move beyond Western samples, the very neat two-dimensional structure becomes more fuzzy and considerably more disorganized. One option to account for this fuzziness is to conclude that we are just dealing with really noisy and messy data. If this were the case, then the disturbances that create these messy empirical data should be random. Yet, in Chapter 8 I review a number of theoretical models and empirical data, demonstrating that the deviations from the neat two-dimensional structure are (a) systematic and (b) can be explained by economic and social conditions within which individuals are living. In other words, the context matters. As surprising as they might be, such findings are compatible with definitions of traits (and values) that many researchers take for granted. In Chapter 9, I extend this argument by examining theories that have proposed societal differences in traits and values. Chapters 8 and 9 therefore bring the focus squarely back to the importance of the economic, social, and cultural context in shaping both the structure and relative importance of traits and values.

The final chapters then return to the dynamic aspects of personality. First, I examine how traits and values develop over the lifetime of an individual. In Chapter 10, I discuss various developmental models and the importance of different periods in a person's life during which traits and values are formed and how they may be elaborated in the minds of individuals over time. In Chapter 11, I turn to the longer-term evolutionary trends. I discuss the most likely models of how evolutionary processes may have shaped the development of traits (and values). Much of this chapter is speculative because of the long-term nature of these events and the limited information that we currently can glean from examining traces in the genetic record. In the final chapter, I consider some of the implications of the overall model for various important issues in modern societies.

Having given a broad overview of the content, you should have a good feeling for what I am trying to argue. For those of you with little patience and a limited attention span (which is me most of the time), here is a synopsis of my main argument.

A SYNOPSIS IN 496 WORDS

Values and personality traits are systematically linked, and their relationship can be mapped out in a two-dimensional space. The organization of values and traits reflects a deeper underlying architecture that develops through a complex interplay between genetic and environmental variables. At the minimum, two major systems are operating. A broad behavioral approach versus avoidance system organizes the first dimension, separating Openness to change values, Openness, and Extraversion traits at one end from Conservatism values, Conscientiousness, and Neuroticism traits on the other end. The second dimension is organized by two different modes of cooperation: an instrumental control orientation and a trust-based orientation, with Extraversion, Conscientiousness, and Self-Enhancement values located at the instrumental end and Agreeableness and Self-Transcendence values at the trust-based end. The interplay between these two dimensions is shaped by the environment, which means that both the relative organization of values and traits and their relative importance and prevalence are conditional on the context within which people are living. This idealized organization varies across contexts because the underlying biological dynamics are expressed differently depending on specific environmental conditions. Most importantly, economic resources, climatic challenges, and to some extent parasite stress shape both the importance and spatial organization of values and traits across different human populations. In less threatening and more supportive environments, traits and values become more clearly separated. In contrast, in threatening and deprived environments, the two-dimensional structure contracts and is likely to exhibit a more narrowly organized self versus other differentiation that is adaptive for survival in these environmental conditions. Similarly, individuals who are embedded in more traditional group structures that do not allow an easy transition across groups require different social skills than do individuals who have free choice of group membership. Hence, socially focused traits and values are likely to diverge across the spectrum of social living conditions. What explains this relative plasticity is that it is evolutionary adaptive. Both early childhood and early adulthood are crucial time windows when the environment is shaping the overall personality (traits and values) of an individual. But personality shows plasticity until late in life, especially when new roles and social contexts require different behavioral responses. This plasticity reflects the adaptive nature of traits and values. Personality traits and values are ultimately influenced by an incredibly large number of genes that each contribute a tiny amount to what a person is like, while the environment continues to shape these contributions in complex and manifold ways. Simple gene-culture coevolutionary models are unlikely to explain these patterns. The most realistic options are complex interactions in which multiple genes regulate an array of brain mechanisms via a multitude of neurotransmitter, hormones, and ion channels that are directly and indirectly transformed by the environment. Another plausible evolutionary process is so-called condition-dependent selection, in which personality traits and values are conditioned on the state of the body. It is at this more complex level that evolution may have exerted some long-lasting influence on who we became as humans.

WHY SHOULD WE CARE?

In the first part of the book I argue that values and personality traits are actually more similar than often acknowledged. I also try to convince you that there are common underlying processes that explain the structure of values and traits that has been documented in many survey studies around the world. You may ask yourself, why is this all important? Why should we care?

Given the widespread use of survey data as the basis for making all sorts of decisions that dramatically affect individuals – think of personnel selection and recruitment, judging the fitness of individuals to stand trial, clinical and educational assessments – we need to carefully evaluate how we measure individual differences in traits and values. As I discuss in detail later, the structure of personality (both values and traits) is not as stable as we often assume – the environmental context fundamentally shapes what we are like and how we see ourselves. This important insight challenges some of the widespread practices of assuming that there is ONE personality model that fits all.

Moving from individuals to populations, much policy decision making is based on drawing inferences from large-scale surveys of needs, goals, and attitudes; it is also driven by the assumption that there is one type of policy that is effective for all of us. The day-to-day activities of individuals are reflections of traits and values that a person holds, but who we are is shaped by our environment. This also has implications for social and economic policies that aim to benefit individuals and communities. Can we always assume that people will benefit equally from well-intentioned interventions? This is particularly important in the context of widespread migration, be it voluntary (e.g., student exchanges, business and labor migration) or involuntary (refugees and mass displacement of populations due to war or environmental disasters). I sketch out some implications of the current integrated person model in the final chapter.

More generally, personality is linked to so many other processes that are of importance for psychologists, managers, educators, social workers, politicians, policy makers, and doctors. Just think of attempts to increase or decrease fertility rates, reduce aggression, foster or harness creativity, improve medical compliance, increase achievement orientation in staff and students, facilitate cooperation among different individuals and human communities, or understand political behavior (with all its complexities as shown in the most recent political events unfolding across many nations). All these attempts ultimately link back to who we are – our personalities and values. To address these issues effectively, we need to reckon with the human psyche's potentials and limitations that are built on the interaction between biological systems and environmental influences.

Most fundamentally, I bring these literatures together in order to provide insights into human nature – why are we the way we are? How different are people around the world in their values and personality? How similar or different are Tina and Sandra, Han farmers in China, New York bankers, and Yanomami making a living through hunting and gathering in the Amazon? What might explain both the similarities and differences between Tina and Sandra or any other person on this planet? Let us start the journey.

2

Everything Is Change: A Primer to Evolution

In this chapter, I outline the basic ideas of evolution. Evolution is simultaneously a simple and complex concept. It is simply about explaining change, but what changes and the why, the how, and the how fast of that process can be quite complex. Therefore, I present a brief overview of the fundamentals of biological evolution as first formally proposed by Darwin. I next describe how culture may also be a target and source of evolutionary processes and then turn to modern extensions that bring both biological and cultural evolution together.

FUNDAMENTALS OF EVOLUTION: VARIATION, COMPETITION, INHERITANCE, AND FITNESS

Darwin's theory has three major elements: variation, competition, and inheritance (Brown, Dickins, Sear, & Laland, 2011; Laland & Brown, 2011; Mesoudi, 2009; Mesoudi, Whiten, & Laland, 2006). First, there is variability between individuals within a species on specific characteristics. For example, some individuals might be more outgoing than others, which results in variation on that particular trait in a population. Second, there is competition between individuals over limited resources such as food, nesting space, or mating partners. In a given environment not all individuals will have equal chances to survive or reproduce. The variability in some characteristics of an individual will increase or decrease the chances of survival or reproduction in this competitive struggle. For example, outgoing individuals may be more likely to find novel food resources or meet more potential mates. Therefore, they may be more likely to survive and produce offspring. The last element of evolution is inheritance: offspring inherit their characteristics from their parents. More outgoing parents may have more kids with those same outgoing characteristics on average. The important point is that the traits that facilitated this success are passed on to the next generation.

These three elements are neutral unless they are linked to the idea of fitness – the success of a specific type of behavior in an environment in surviving and thriving (e.g., producing offspring that themselves survive and thrive). Applied to our simple example, individuals who are more outgoing in a specific environment – say, modern Western society – or have a particular set of values that induce them to explore environments are said to be higher in fitness if they are more successful in reproducing and passing on their genes. Over time, these fitness advantages will accumulate, and in each succeeding generation in the same environment that favors a particular trait, that trait will increase in frequency, thereby increasing the fitness of the individual possessing that trait or value. In contrast, in another environment a different type of behavior may be beneficial for survival and thriving. In our example, outgoing behavior was deemed as producing benefit because its rewards were high and the costs low in modern Western society, where the environment is benign and threat free (imposing low costs of injury, poisoning, or some other form of accidental death). In this environment, there are more costs for missing out by not exploring compared to those incurred by exploring one's environment. However, in a more hostile and threatening environment, individuals exploring the environment are at greater risk of being attacked or injured, catching some nasty parasites, or dying from eating poisonous foods. The relative fit of behaviors and survival strategies to a specific environment is therefore important, allowing for change over time through selection. The three elements of variation, competition, and inheritance characterize evolutionary processes; without these elements there would be no evolutionary process. The absence of any one of these features (e.g., traits are not being passed on from parent to the offspring generation, there is no competition between the phenotypes or no variability among individuals) means that no evolutionary processes are operating.

CULTURAL EVOLUTION

These ideas are relatively uncontroversial when applied to biological processes. To what extent are these ideas applicable to social systems and cultures? The field of cross-cultural psychology has documented a large number of psychological processes and traits that differ in their distributions across human populations (Berry et al., 2009; Smith, Fischer, Vignoles, & Bond, 2013). Variation in human populations is inherent to the notion of culture itself: this should be a relatively uncontroversial statement. The idea of competition between cultures may be more controversial, yet human history and contemporary politics provide countless examples of competition between different cultural groups. The legacies of colonial domination and intergroup competition in the form of war, and all of the attendant suffering, are the clearest examples of competition between groups and populations that differ in culture and social organization. The last element of evolution – inheritance – is also quite unproblematic. Individuals are socialized into their cultural system – that is, they learn appropriate cultural norms, beliefs, and attitudes from their parents and elders. Culture is then passed on from generation to generation, a clear form of inheritance.

As already discussed, the relative fit of a biological or cultural phenotype to an environment is a key criterion for survival. How well is a phenotype adapted to a specific context? What are the optimal strategies and behaviors in a specific environment? Human behavioral ecologists differentiate between current levels of *adaptiveness* and evolutionary levels of *adaptation* (Laland & Brown, 2011). The former is an indicator of the functionality of a behavior or strategy in increasing reproductive success in current environmental conditions. The latter is a behavior or strategy that has been selected for its effectiveness over evolutionary time. These two dimensions of time scale and level of adaptiveness give rise to four theoretical possibilities. If a behavior is adaptive today and this behavior has been evolutionary selected, then we are dealing with a *current adaptation*. If the behavior does not increase reproductive success today and has never done so (has not been selected for increasing fitness at any time in evolutionary history), then we are dealing with a *byproduct*. If a behavior was adaptive in the past, but not anymore because of a change in the environment leading to changes in selection pressures, then we are dealing with a *past adaptation*. Finally, a phenotype that today enhances fitness but was not built for this current role by natural selection is called an *exaptation*. These four possibilities are conceptually easy to understand, but distinguishing between what might be an exaptation or current or past adaptation can be quite challenging. Are the motivational characteristics that "made" Tina take part in those religious ceremonies adaptive today, or did such motivational characteristics provide some fitness benefits to her ancestors? Is the way Sandra has organized her desk beneficial for her fitness, or is it a byproduct of some other feature that helped her ancestors survive? Much of this debate centers around selection pressures and adaptive responses in earlier times that are difficult to reconstruct.

IS THERE A SINGLE ENVIRONMENT OF EVOLUTIONARY ADAPTIVENESS?

A central assumption of evolutionary psychology is that our brain mechanisms that characterize and drive contemporary responses to the environment were set during the Pleistocene, when modern humans (Homo sapiens) first emerged as a biological species in the larger family of hominids (including such illustrious but extinct relatives with names like *australopithecus* or *ardipithecus*) (Cosmides & Tooby, 1989, 1992). According to this classic evolutionary psychology approach, our capacities and psychological make-up have not changed since the Pleistocene because of the slow process of biological evolution, which occurs when our genes change due to random mutations. There has simply not been enough time for further evolution (read: genetic change) to take place between the time we first evolved in the African savannah and modern times. Thus, any changes that we see between human populations today are evidence of cultural evolution, which is happening at a much faster rate.

This classic argument has been challenged by recent evidence suggesting that indeed there have been changes in our genetic make-up in more recent human history as humans both adapted to novel environments, and adopted certain cultural practices. Evolutionary processes are most likely to affect changes in the ability to survive (e.g., to adapt to local conditions and secure sufficient food) and produce healthy offspring. Not surprisingly, some of these changes have to do with the ecological environments into which humans moved. For example, adaptations to colder or drier climates seem to have left their genetic mark on populations that moved into these environments (Hancock et al., 2010). There are many small but cumulative and functional mutations in genes distributed across the genome that help humans regulate body temperature and metabolize food in colder climates. These polymorphisms are differentially distributed around the world. For example, Naukan Yup'ik, Yakut, and Maritime Chukchee populations in the far northern regions of Russia show variations in the pathways associated with key processes for metabolizing sugars and fat to create energy.

Other genetic differences have developed as a result of cultural changes since the Pleistocene. The most famous one is the emergence of lactose tolerance in those populations in Europe and parts of Africa and Asia that relied on cattle (Beja-Pereira et al., 2003; Durham, 1991). All humans have an ability to digest milk as infants, but this ability disappears after babies are weaned off breast milk; thus, adult humans were historically unable to digest milk. With the advance of agriculture and the domestication of cattle, however, milk became available for consumption. Humans who had the ability to digest milk and benefit from its calories, nutrients (especially calcium, which helps prevent the breakdown of Vitamin D), and water content were more likely to prosper and their offspring would have a higher chance to survive. These mutations spread with incredible speed across populations that had domesticated cattle. Yet the functional mechanisms were different across populations, likely driven by the water content of milk in dry African climates (where water in the milk is a valuable source of hydration) and the richness of nutrients (especially calcium that is linked to Vitamin D processing) in milk in relatively dark (and therefore vitamin D deficient) northern European climates (Ingram, Liebert, & Swallow, 2012). Another example of a recent genetic change is the ability to digest cereals, roots, and tubers (Hancock et al., 2010). Populations such as the Yoruba, Papuans, Melanesians, and Karitiana living in environments with scarcer food sources where roots and tubers become important sources of nutrition have higher frequencies of polymorphisms that help them break down the starch and sucrose in tubers and extract more folate from them (which is an important vitamin, but is not highly concentrated in roots and tubers). All these findings suggest that there is no single environment of evolutionary adaptiveness, at least not for basic body functions. However, many genes have so-called pleiotropic effects; that is, they have multiple effects throughout the body and central nervous system, which may influence those behavioral aspects of the person that we typically call personality (Penke, Denissen, & Miller, 2007).

GENE-CULTURE COEVOLUTION

These examples suggest a strong interplay between biological and cultural processes, often summarized under the label of gene-culture coevolution (Richerson & Boyd, 2005) or niche construction theory (Flynn, Laland, Kendal, & Kendal, 2013; Laland, Odling-Smee, & Feldman, 2000). These models assume that genes and culture have jointly coevolved to provide gene carriers (individuals) with the necessary fitness-enhancing traits to survive in a crafted social and ecological environment. Our actions change our environment, and this change then sets in motion feedback mechanisms that have an impact on our genes. Therefore, the relative fitness of individuals carrying certain versions of a gene (called alleles) will differ from the fitness of an individual who carries the complementary set of genes, changes that are the long-term consequences of the collective actions of our ancestors that changed the selection pressures in our cultural environment. For example, I am able to digest milk because my ancestors a few millennia ago started farming cows and consuming milk. Some of my ancestors, after they were weaned from breast milk, had a strange new mutation that allowed them to digest the specific sugar present in cow's milk without suffering nausea or diarrhea. Because my ancestors were located in the far northern part of Europe, the changes in the vitamin D processing then gave those with that copy of the mutant gene a slight advantage: it allowed them to obtain a vital vitamin in an environment where it was rare due to reduced sunlight (which the body needs to produce the vitamin). Being a little bit healthier than others, my ancestors with this advantageous mutation in their genes then had a better chance to have more offspring and prosper. They also may have selected for breeding those cows with the tastiest milk, setting into motion a second selective process (Beja-Pereira et al., 2003), by which we humans then started breeding those cows that were most useful for us (those that produced the largest amount of milk with the best features).

THE PROBLEM OF TIME SCALES

These processes of gene-culture coevolution took place over millennia. Yet, time scales of change in biological and cultural systems do vary. Biological evolution is tied to the passing of genetic information from parents to offspring. The relatively slow processes of mutation and recombination of alleles passed from parents to offspring only take effect in a population over many, many generations. Functionally advantageous mutations are relatively rare, and the time scales in which such beneficial mutations occur and take hold in a population are vast. Yet, biological evolutionary processes can also occur at a much faster time scale. Seasonal changes in weather that affect food supply with immediate effects on phenotypes have been documented across a number of species. The best-known example is the change in beak size and structure in Darwin finches. In dry seasons, birds with harder and longer beaks are more likely to secure enough food (e.g., by collecting and cracking harder seeds that can withstand dry weather) to feed their offspring; in contrast, fledglings of birds with beaks better adapted to eat soft fruits or insects are less likely to survive. Therefore, already after one season, mean beak sizes have shifted, resulting in rapid evolutionary change over a few generations (Grant & Grant, 2006). In some extreme cases, such evolutionary pressures can change the phenotype of a population overnight.

A severe snow storm in spring can kill off most returning migratory birds, and only those birds with a greater cold tolerance may survive and pass their genes on to the next generation (Zuk, 2013). Hence, biological evolution operates on both historical and seasonal time scales.

As mentioned, cultural evolution is thought to occur more rapidly than biological evolution. Information can be passed on from individual to individual and from population to population within and across generations. Therefore, novel information can theoretically spread very rapidly and lead to much faster evolutionary changes than biological evolution, which is dependent on genetic transmission. At the same time, inertia in cultural systems is also relatively strong. New technology may not emerge for long periods of time, and cultural patterns may remain stable over hundreds or even thousands of years. Much traditional technology, such as hunting, fishing, or agricultural methods, has remained largely unchanged in many societies around the world. Yet, the rapid change brought about by modern technology shows that cultural evolution has a tremendous potential for much more rapid change than does biological evolution. When I grew up, there was a single phone in my village, and the main form of long-distance communication was to write letters. Writing this book on a laptop that is connected to the Internet, I constantly have to restrain myself from talking to friends on Facebook or other social media platforms. The change in cultural evolution within one generation is phenomenally large and cannot be matched by natural selection and genetic changes. These differing time scales will not concern us initially, but we will need to grapple with time once our basic framework is in place and we examine how personality develops and changes.

THE RANDOM GAME OF GENES

One common assumption is that genes form the genotype, which then determines the phenotype (the behaviors expressed by the genes; e.g., what a person is like). The corollary is that genotypes are selected so that changes in selected genes result in phenotypes that survive better in a new environment. Yet, the reality is much more complex, and many random and systematic processes exist that will be important to explore in the context of our intellectual voyage. First, we have to consider the null hypothesis that changes in a gene pool of a population are random. This is often called *genetic drift*, meaning that slow changes in the frequency of a particular gene version are due to accidental events that kill people or do not allow them to reproduce. This genetic drift process is absolutely random and does not mean that the genes of the survivors are in any way better adapted than of those who did not make it. Think of a volcanic eruption that occurs on a small island, near where all the villagers are working on a field. Let's assume that the eruption occurred rapidly and without warning. Some individuals who were by chance farther away from the eruption might survive, whereas those who were closer might die. There is no adaptive reason why some people just happened to work the part of the field closer to the top of the volcano and therefore did not survive.

Second, we need to consider the important process of bottlenecks and founder effects. Imagine that a small group of people splits off from a much larger population and moves to a new environment. This split could be completely random: think of a convoy of canoes that travels across the ocean, and by an accidental force of nature, one of the canoes is blown off course and stranded on an uninhabited island. The men and women who were on the canoe settle on the island and start reproducing. Because the small group is unlikely to represent the diversity of the overall population, its genetic variability is greatly reduced. This is the proverbial bottleneck, which is caused by only a few people being in the canoe and bringing their genes with them. The particular genes that are present in this new population were clearly not selected biologically or culturally. Furthermore, given the much smaller population, the probability of random processes acting on the gene pool (random accidents or recombinations that decrease certain gene variants in the small population) is even larger, increasing the chances of genetic drift. *Bottlenecks* do not need to involve migration – they are operative whenever the population is shrinking for at least one generation. This reduction in population could be due to war, natural disasters, famines, or epidemics.

As the small group of survivors on the deserted island start reproducing, their genes are copied and transferred to the next generation. Essentially, the individuals with their particular set of genes now become the founders of a new population. The characteristics of the original members now are much more prevalent in this population. This is often called a *founder effect*.

In the bottleneck and founder effect cases, the frequency of certain genes may have nothing to do with adaptiveness to the environment. Yet, it might be tempting to observe a certain distribution of genes in a particular environment and conclude that some selection events were happening. This is one of the challenges that we need to consider when discussing genes and behaviors across different populations around the world. Modern genetics in fact has ways of leaving genetic signatures that indicate whether and even when some selection may have occurred. These insights will be important later when we examine selection possibilities on human personality traits.

EPIGENETICS AND GENE-ENVIRONMENT INTERACTIONS

Genes are important. Yet, biologists are finding more and more evidence that genes can be switched on and off by events in a person's lifetime, driven by complex constellations of environmental factors. This general process is called *epigenetics* – which in a narrow technical sense means that the function of a gene is changed without altering its actual genetic information. Even though an individual may carry a particular gene that produces a particular protein or determines how effectively synapses are working in the brain, it may be silenced by complex biochemical processes. Even more surprising, the state that genes are in (whether they are switched on or off) can be transferred across generations. This means that, if your grandparents encountered some particular environmental stressor (let's say they were facing starvation), their genes might be set in a particular activation pattern (some are set to on, some are set to off), which then is transferred to your parents and from your parents to you (Heijmans et al., 2008). Therefore, some researchers have started to define epigenetics in this broader sense, which refers to any kind of heritable change that is not transmitted via genes.

Epigenetics in both the narrow and the wider definition implies that the environment has an important influence on who we are not just by providing some selection pressure but also by shaping how genes are operating. These insights have started to emerge in the last few decades, mostly from research focusing on health problems, including obesity and mental health. It is now time to consider how these processes may influence personality and values. Given the current state of our understanding of how these epigenetic processes work and how they may influence more complex behaviors, this is largely uncharted territory. But in this book I show some empirical examples that suggest that these processes warrant closer scrutiny and empirical investigation. Epigenetics will be of importance in the later sections of this book.

SUMMARY

This chapter covered a number of fundamental concepts and ideas that we will need to evaluate traits and values. Evolutionary processes require variation, competition, and the heritability of specific personality phenotypes.

Selection works by favoring certain phenotypes that are a better fit to a particular environment. Therefore, adaptiveness to the local environment is a key idea that we will come back to repeatedly. At the same time, genetic differences do not necessarily mean that selection has taken place. Purely random processes such as genetic drift, bottlenecks, and founder effects may account for those differences. When selection is indeed taking place, evolutionary processes can involve both genetic and social changes, and most likely a combination of both via epigenetic pathways and gene-culture coevolution. The time scales for what is seen as adaptive and how fast different types of evolutionary processes take place are nevertheless complex issues. But then, personality is complex, and why should the mechanisms that have led to different personalities be any different or simpler?

Now that we have a basic grasp of evolutionary basics, it is time to look at personality traits and human values and how they are studied in psychology.

3

The Big Five Personality Traits and Human Values

How would you describe yourself or somebody else? What characteristics come to mind when you want to describe a friend to a stranger? To provide some context for what I discuss in this chapter, it would be helpful if you would take a pen and write down the characteristics that come to mind when describing yourself or a person close to you. Be as specific or general as you wish, but try to come up with an accurate profile.

Done? Now, let's have a look at that profile. You probably have noticed that you can use many different types of characteristics or categories for describing yourself or others. Table 3.1 provides an overview of categories that people may use when describing a person (Nel et al., 2012; Valchev, van de Vijver, Nel, Rothmann, & Meiring, 2013). There is a bit of overlap and similarity between these categories, and their boundaries might be fluid. For example, being helpful could be a trait, a virtue, and a value. The pioneers of modern psychology such as Gordon Allport thought that both values and personality traits are important for the description of individuals. Yet, over the course of the second half of the last century, personality and values have been studied in largely independent lines of research. Attitudes, emotions, and competencies also received a significant amount of attention, but they were also studied largely independently and in isolated research lines. Other areas such as virtues and vices received considerably less attention. In this book, I focus on values and personality traits for several theoretical reasons that I discuss in the chapters to come. Additionally, a very important pragmatic reason for this exclusive focus is that these two areas have attracted more attention than other elements in the description of humans. Importantly, standardized tests are available that have been applied to large numbers of human groups around the world. If we want to describe what people

TABLE 3.1 *General Categories Used for Describing Individuals*

Content	Definition	Example
Trait	Habitual patterns of behavior, thought, and emotion that are relatively stable over time and situations, similar to a habit	Helpful, outgoing
Virtue	Positive trait or quality that is consensually judged as good and highly valued in a social community, has moral connotations (moral excellence)	Courage, tolerant
Behavior	Specific action taken by an individual in a specific environment	Smiles, shakes the hand
Values	Abstract beliefs about desirable goals or modes of conduct that transcend specific situations; serve as guides for selecting and evaluating behavior, people, and events; and can be systematically ordered according to their relative importance	Helpful, independence, clean, courage
Vices	Behavior, trait, or habit that is considered negatively (sinful, defect, unhealthy, etc.) in a social context and community, has strong negative moral connotations (related to sins)	Lazy, vain, angry
Preferences	General statement that somebody likes one object or action more than another	Likes, enjoys, wants
Attitudes	Evaluative statement of an attitude object, which has cognitive (beliefs and thoughts), affective (feelings and emotions), and behavioral (past behaviors and experiences) components	Likes, believes, thinks
Competencies	Ability of an individual to complete a more complex set of actions	Smart, is able to, intelligent
Emotional state	Temporary state of the emotions and feelings of a person	Is afraid, irritated
Role	Set of behaviors, rights, and obligations that are associated with a specific function that an individual assumes in a social situation	Is like a father, behaves like a boss
Role fitting	Evaluation of whether a person conforms to the expectations of a role	Is not a good mother, is a good teacher
Social identity	An individual's sense of who he or she is based on membership (or subjective feeling of belonging) in a social group or category	Muslim, Brazilian, student
Religious belief	Belief in mythological, supernatural, or spiritual concepts	Believes in God
Social status	Position or rank of an individual within a social group	She is respected, he is our leader

TABLE 3.1 *(continued)*

Content	Definition	Example
Contextualization		
Life domain	Aspect of life when a person shows or expresses some behavior, trait, value, or other content dimension	Helps fellow colleagues
Temporal/ situational	Describes the context, frequency, or duration of a content dimension	When, sometimes, rarely
General relations	Describes the content dimension in relation to other people in general	With other people, is generous to others
Specific relations	Describes the content dimension in relation to specific individuals	With his partner, he is . . .

Note: This table is adapted and extended from Valchev et al. (2013).

are like around the world, we need comparable data. Only comparable data can give us valid information to unpackage differences and similarities in human personality. The issue of comparability will keep us busy in chapters to come (most centrally in Chapter 6).

In this chapter, I start the discussion of values and personality traits, the core variables of interest in this book. As long ago as ancient Greece and imperial China, philosophers and thinkers mused about personality differences between individuals. Back then, influential thinkers such as the forefathers of modern medicine, Galen of Pergamon and Hippocrates, thought that different combinations of bodily fluids shape the personality of a person and lead to different personality types. These typologies in ancient thought have mainly been replaced with a dimensional system in modern psychology. Most research published in psychology today that explicitly mentions personality is based on analysis of trait terms. The study of values has developed in parallel, inspired in particular by the work of sociologists and anthropologists. I discuss the general approach to measuring both traits and values as scientific constructs and their empirical structure as identified in Western samples. The structures described in this chapter are central to my arguments about the functional organization of personality and evolutionary adaptiveness that I will develop later.

To avoid confusion, I use the term "personality" when talking about personality in general, including traits and values. I use "traits" or "personality traits" when talking about the specific research tradition that I introduce in the next section.

PERSONALITY TRAIT DIMENSIONS: THE BASIC PICTURE EMERGING IN WESTERN SAMPLES

Traits can be defined in a number of ways. The most common definitions refer to traits as habitual patterns of behavior, thought, and emotion that are relatively stable over time and situations (e.g., Valchev et al., 2013); as behavioral reaction norms (Dingemanse, Kazem, Réale, & Wright, 2010) that provide "probabilistic descriptions of relatively stable patterns of emotion, motivation, cognition, and behaviour, in response to classes of stimuli that have been present in human cultures over evolutionary time" (DeYoung, 2014); or as a density distribution of personality-relevant states (Fleeson, 2001). What is common to these definitions is the emphasis on the relative consistency of behavioral predispositions to behave in a particular manner across situations. A person is described as helpful (as a trait) if she is likely to consistently behave toward others in ways that help them, to the extent that they are in some form of need. Therefore, a trait captures behavioral dispositions that emerge in reaction to specific types of situations.

The Lexical Hypothesis and the Big Five

Approaches to studying traits are grounded in what has been called the linguistic approach and the **lexical hypothesis** (Goldberg, 1981, 1993). This hypothesis, first formulated in 1884 by Sir Francis Galton (1949), proposes that the most important elements of personality are encoded in language and should be represented by clusters of highly similar words in all human languages. To the extent that a term is particularly important, it should be represented in a single word (or a small number of highly synonymous terms) so that individuals can easily communicate about this trait. In other words, it assumes that if there is an important dimension of personality, then people in their everyday descriptions of each other and themselves should use same clusters of words to refer to the same type of behavioral traits. Let's come back to your descriptions from the beginning of this chapter. Can you identify some general patterns in the words that you use to describe yourself and other people?

Based on the lexical hypothesis, researchers use dictionaries to cull words that might be used to describe individuals. These words are then given to participants to describe themselves or others, typically by using some form of a rating scale. This scale requires the respondent to say whether the statement applies to him or her (1 "Does not apply at all" to 7 "Completely applies") or whether the respondent agrees with the statement (1 "Totally disagree" to 5 "Totally agree"). These ratings are then typically analyzed with statistical techniques such as **factor analysis** – a technique to reduce complexity in large amounts of data. If a large number of respondents answer in a similar way to two different statements (e.g., assigning a "1" both to "I like to meet new people" and "I often am the center of the party"), then these two items will likely emerge on the same statistical factor. The assumption is that there is an underlying "causal" variable that leads the respondent to respond in the same way to these two statements.

Based on these techniques and reasoning, the fundamental dimensions of human personality traits should be identifiable if we give large numbers of trait terms to individuals and then submit the responses to factor analysis (Ashton, Lee, & Goldberg, 2004; Goldberg, 1993). Goldberg (1993) reviewed these initial studies and pointed out that researchers used as few as 60 English words. These lists of words were then administered to large samples of participants to identify core dimensions of personality traits. A few problems should become apparent with this methodology: it may make a difference if (a) we restrict the number and diversity of words that we present to individuals; (b) the context of description (the contextualization dimensions in Table 3.1) may vary; and (c) the language of the words may make a difference. In this chapter, I mainly deal with the first problem. Chapters 7 and 8 deal with issues of context and language.

The most widely accepted and used framework for studying personality in this tradition today is what is variably known as the five-factor model (FFM), Big5, or Big Five (Goldberg, 1981; McCrae & John, 1992). There are various tests that have been developed in this tradition, including the NEO Personality Inventory-Revised (NEO-PI-R) (Costa & McCrae, 1992), the Big Five Inventory (BFI; Benet-Martínez & John, 1998; Soto & John, 2009), and freely available open-source versions such as the International Personality Item Pool (IPIP; Goldberg et al., 2006). All these instruments typically ask respondents to report whether they agree with statements, such as "I see myself as someone who . . . tends to be lazy" or "is outgoing, sociable" (e.g., from the BFI); "I am not a worrier" or "I sometimes lose interest if people talk about abstract matters" (NEO-PI-R); or "I am the life of the party" or "I get stressed out easily" (IPIP).

The various approaches to measuring these five factors typically agree on the number of factors (even though they use different labels). There is also consensus on some but not all of the content of these factors across the different questionnaires. In this book, I use the terminology of the NEO-PI-R developed by Robert (Jeff) McCrae and Paul Costa. The basic five major dimensions in this model are called Conscientiousness (C), Agreeableness (A), Neuroticism (N), Openness (O), and Extraversion (E). The NEO-PI was developed from an earlier three-dimensional model of personality first developed by Hans Eysenck (Barrett, Petrides, Eysenck, & Eysenck, 1998; Eysenck & Eysenck, 1965).[1] Goldberg's (1993) work was instrumental in increasing the acceptance of the five factors among the larger academic community. It is reassuring that different approaches and questionnaires converge around a similar number and content of factors across a number of independent studies.

Essentially, these trait dimensions capture personality as it is embedded in everyday discourse about people. They capture the basic dimensions that people use to communicate important information about themselves or others. Individuals high in Extraversion are typically described as active, assertive, energetic, enthusiastic, outgoing, and talkative. Persons high in Agreeableness are appreciative, forgiving, generous, kind, sympathetic, and trusting. Conscientious individuals are thought to be efficient, organized, planful, reliable, responsible, and thorough. Anxious, self-pitying, tense, touchy, unstable, and worrying are adjectives used to describe individuals high in Neuroticism. Finally, people high in Openness are described as artistic, curious, imaginative, insightful, original and as having wide interests. These are mainly positive terms (except for Neuroticism), but each personality dimension also has a number of polar opposite (potentially negative) trait adjectives.

The Hierarchical Organization: Five Splitting into Ten

These five traits form the most important level of hierarchically organized behavioral classification. Colin De Young and colleagues (DeYoung, Quilty, & Peterson, 2007) were the first to systematically classify the lower-order organization underneath these major five traits. They found consistent covariation within each trait that led them to argue that there are intermediate factors existing within the Big Five, which they called aspects. These aspects differentiate among major motivational and behavioral differences within the overall factors. For example, Neuroticism has a withdrawal and a

[1] Two of the dimensions proposed by Eysenck have clear equivalents in the FFM (Neuroticism and Extraversion). The third trait was called Psychoticism and has been considerably more controversial. Focusing just on the empirical relationships, Costa and McCrae (1992) argue that Psychoticism is the opposite of Conscientiousness and Agreeableness.

volatility aspect. Withdrawal refers to inhibiting behavioral impulses, anxiety, depression, self-consciousness, and vulnerability, whereas the volatility aspect might also be related to emotional lability, irritability, and reactive aggression (e.g., anger outbursts and violence when provoked). These aspects help structure the often complex content of the overarching trait. There is also a biological rationale for separating these traits and aspects, which I review in Chapter 6. Table 3.2 shows the breakdown of the five traits into the respective trait aspects and gives examples for each aspect.

Interpreting the Five Factor: The HEXACO Approach

An interesting discussion has emerged around how to best interpret these five traits. As we have already seen, the overall traits are complex, and trait aspects might exist within each higher-order trait. A large-scale review of lexical analyses by Ashton, Lee and Goldberg (2004) suggested that three of the factors they found (C, E, and O) were straight replications of the traditional Big Five. However, Agreeableness and Neuroticism showed a somewhat different organization, which motivated Ashton and colleagues to develop an alternative model that they called HEXACO. I will break down the acronym later, after I describe the major changes in their model. They renamed Neuroticism as Emotionality. Emotionality in the HEXACO captures emotional reactivity rather than emotional instability and excludes Anger (part of Neuroticism in the Big Five; see volatility) and includes sentimentality (part of Agreeableness in the Big Five). Agreeableness in their model excludes sentimentality, but includes lack of anger. Therefore, the Agreeableness factor in the HEXACO model is less emotionally charged and captures the more interpersonal aspects of patience, tolerance, and peacefulness. They also derived another factor from Agreeableness called Honesty-Humility: this factor is characterized by personality traits related to sincerity, fairness, being unpretentious, and lacking greed. As we will see in the next chapter, Agreeableness and Honesty-Humility are often closely related, even within the original data set. The six factors provide the base for the acronym H(onesty-Humility) E(motionality) eX(traversion) A(greeableness) C(onscientiousness) O(penness).

Comparing the two models, the HEXACO makes more fine-grained differentiations for two of the five traits (Neuroticism and Agreeableness) and rotates them slightly in their meaning and orientation; it retains the three other traits of the Big Five without further changes in meaning. Therefore, the Big Five and HEXACO models are relatively compatible with each other.

TABLE 3.2 *Relation of the Content of the Big Five, the Five Factor Aspects, and the HEXACO*

	Big Five	**Five-Factor Aspects (DeYoung et al., 2007)**	**HEXACO**
C	organized, systematic, efficient, precise, thorough, practical *versus* careless, sloppy, absent-minded, haphazard, disorderly, unreliable	**Industriousness**: carry out my plans, finish what I start; waste my time (reversed); find it difficult to get down to work (reversed) **Orderliness**: like order; leave my belongings around (reversed); keep things tidy, follow a schedule	organized, disciplined, diligent, careful, thorough, precise *versus* sloppy, negligent, reckless, lazy, irresponsible, absent-minded
E	talkative, extraverted, sociable, assertive, enthusiastic, verbal *versus* withdrawn, silent, introverted, shy, reserved, inhibited	**Enthusiasm**: make friends easily; warm up quickly to others; have a lot of fun; keep others at a distance (reversed); reveal little about myself (reversed) **Assertiveness**: take charge, have a strong personality; know how to captivate people, wait for others to lead the way (reversed)	outgoing, lively, extroverted, sociable, talkative, cheerful, active *versus* shy, passive, withdrawn, introverted, quiet, reserved
O	intellectual, complex, philosophical, innovative, unconventional *versus* simple, conventional, uninquisitive, unintelligent, shallow	**Intellect**: am quick to understand things; have difficulty understanding abstract ideas (reversed); can handle a lot of information; like to solve complex problems **Openness**: enjoy the beauty of nature; love to reflect on things; need a creative outlet; get deeply immersed in music	intellectual, creative, unconventional, innovative, ironic *versus* shallow, unimaginative, conventional

TABLE 3.2 *(continued)*

	Big Five	Five-Factor Aspects (DeYoung et al., 2007)	HEXACO
N/Emotionality	relaxed, unemotional, easy-going, unexcitable *versus* moody, jealous, possessive, anxious, touchy, high-strung	**Withdrawal**: am filled with doubts about things; feel threatened easily; worry about things; seldom feel blue (reversed) **Volatility**: get angry/upset easily; rarely get irritated (reversed); change my mood often	emotional, oversensitive, sentimental, fearful, anxious, vulnerable *versus* brave, tough, independent, self-assured, stable
A	sympathetic, kind, warm, cooperative, sincere, compassionate *versus* cold, harsh, rude, rough, antagonistic, callous	**Compassion**: feel others' emotions, inquire about others' well-being, am not interested in other people's problems (reversed) **Politeness**: respect authority; hate to seem pushy, insult people (reversed); love a good fight (reversed); believe that I am better than others (reversed)	patient, tolerant, peaceful, mild, agreeable, lenient, gentle *versus* ill-tempered, quarrelsome, stubborn, choleric
H	-	-	sincere, honest, faithful/loyal, modest/unassuming, fair-minded *versus* sly, deceitful, greedy, pretentious, hypocritical, boastful, pompous

Note: Content is adapted from various sources (Ashton & Lee, 2007, 2008; DeYoung et al., 2007).

Summary of the Big Five

The lexical trait approach provides a convenient descriptive theory of personality language. The five main factors capture salient distinctions that laypeople make in describing themselves and others. One important caveat is that the trait structure captures patterns of covariation of traits **across** individuals in a sample; these patterns do not necessarily mean that the traits are organized in the same way within individuals. These dimensions are a simplification of the richer nuances of how individuals differ from each other, but they are a convenient and relatively parsimonious summary of key differences between individuals. The Big Five model is a classification of major dimensions of interindividual differences, as perceived and used by individuals when describing themselves and others.

If you described yourself or another person earlier, you can check Table 3.2 to see how your self-description matches onto the major positive and negative trait adjectives for each model and dimension. It also shows how the HEXACO and Big Five differ in their rotation of Agreeableness and Neuroticism.

BASIC HUMAN VALUES

Values have long been of interest to philosophers and theologians (Maio, 2016). This historical interest across disciplines has led to a diversity of definitions and approaches, which nevertheless share a common core. Schwartz and Bilsky (1987) examined and summarized the manifold theoretical approaches and definitions of values, and distilled these into this overall definition: values are beliefs that refer to desirable goals or modes of conduct; transcend specific situations; serve as guides for selecting and evaluating behavior, people and events; and can be systematically ordered according to their relative importance. In other words, values can be seen as motivational orientations that capture general life goals and instructions of how to achieve them (Fischer, 2013). Gordon Allport (1955), one of the grandfathers of psychology, claimed that the study of values was important for understanding the personality of individuals in his seminal book, *Becoming*. Influential pioneers of personality research adopted definitions of personality that encompassed values as important aspects of personality (Guilford, 1959). Yet, the study of values has developed separately from personality, and today there is little reference to values in standard texts on personality (Bilsky & Schwartz, 1994; Nilsson, 2014). A major and obvious difference between the two is that researchers typically measure traits in terms of how people typically behave, whereas values are typically rated in terms of their importance. Therefore, trait dimensions capture perceived behavioral consistencies, whereas values capture what people deem important in their lives.

Much of the modern research on values within psychology has its roots in the classic sociological work by Kluckhohn and Strodtbeck (1961) that examined value orientations in small communities in the United States during the late 1940s and 1950s. This early work had some strong influence on later cross-national studies such as Hofstede's (1980) work. Hofstede studied value-like statements among IBM employees in the 1960s and 1970s and developed his famous four-dimensional model of cultural values. The dimension that has attracted most attention is individualism-collectivism, which captures the importance of the individual versus the group. Yet, Hofstede's work was at the societal level and did not provide much insight into the values of individuals.

From an individual perspective, Milton Rokeach's (1968, 1973) work on values in sociology had the largest influence on empirical value research in psychology. Rokeach's work was based on the lexical hypothesis, which I just discussed. He selected 36 values from a variation of the original trait list of nearly 18,000 words first compiled in 1936 (Allport & Odbert, 1936), the same list that formed the basis of what would become the Big Five. Rokeach only selected positive trait terms and dropped negatively valued terms. As should be obvious, the selection of only 36 values raises questions about the representativeness of any value instrument. Yet, Rokeach's work provided the foundation for value research in psychology and has influenced a large number of researchers studying value problems, even though it never clearly mapped out the larger structure of the value domain. To address this gap, Shalom Schwartz iteratively developed a descriptive theory of human values (Schwartz & Bilsky, 1987). Schwartz's major achievement was to enable a clearer understanding of the organization of values. This work at the individual level provided the much-needed structure that could complement and extend Hofstede's work at the societal level.

As with personality traits reviewed earlier, values as conscious endorsements of motivational goals can be organized into a simple structure. The simplest way of organizing values is a two-dimensional representation in a single plane in which values are ordered in terms of their compatibilities and conflicts with each other (Fontaine, Poortinga, Delbeke, & Schwartz, 2008). Schwartz (1992) used a circle to represent the organization of values, but other spatial representations of the same value space are also possible (Gouveia, Milfont, & Guerra, 2014). A major analytical choice in analyzing the data compared to personality research is that Schwartz opted for a multidimensional solution that explicitly represents and describes the relationship of values to each other. The most common technique in this family of statistical methods is **multidimensional scaling** (MDS; Borg & Groenen, 2003, 2005). It reduces the complexity of a large set of data and displays the similarity of objects within a data set visually. The closer that two objects are to each other within a multidimensional space, the more similar they are. Conversely, the farther away two objects are from each other, the more dissimilar they are. In contrast, the trait approach, which attempts to find independent dimensions of traits via factor analysis, is less concerned with the relative position of traits to each other. The advantages of MDS are that it accommodates a much larger diversity of data, provides solutions that do not assume the independence of dimensions, and is able to display correlated dimensions.

As with the trait approaches, Schwartz's research had to use a small set of values to avoid fatiguing his participants. He drew on Rokeach's work, a Chinese inventory of values (Bond, 1988), and values proposed in discussions with religious leaders to create a list of 57 value terms. Over a series of studies, he developed this list into a structure that he described as a near-universal structure of human values (Schwartz, 1992).

Schwartz's Value Theory

The two-dimensional structure of values is projected into a plane that can be subdivided in a number of different ways. Originally, Schwartz separated values into 10 value types that are arranged into four higher-order value clusters that form the endpoints of two theoretical dimensions. These two theoretical dimensions are organized in a circular structure. From the beginning, Schwartz emphasized that different distinctions could be made within the same value space. More recently, both he and his colleagues created both finer and coarser distinctions and also discussed rotations of the major orientations of the value space (Fontaine et al., 2008; Schwartz et al., 2012). In this section, I focus on Schwartz's original organization into 10 value types along two major dimensions because it has attracted the most attention by researchers to date and there is now relatively strong empirical support for this particular division of the value space (Maio, 2010).

Figure 3.1 shows the classic circle of value compatibilities and conflicts. Starting with the top left of the circle, Universalism (UN) values are motivations to understand, appreciate, tolerate, and protect the welfare of all people and nature. Moving around in a clockwise fashion, Benevolence (BE)

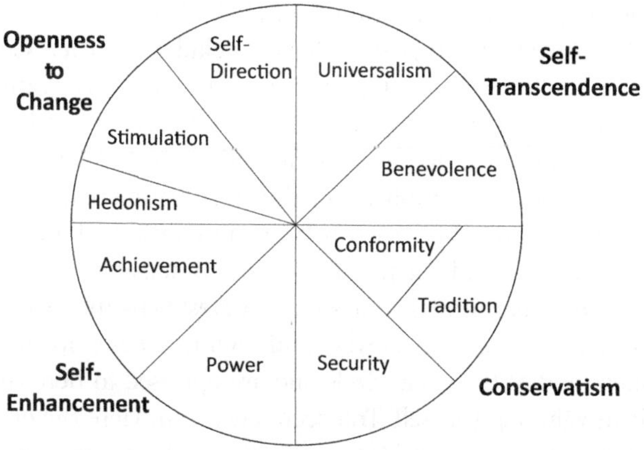

FIGURE 3.1 The prototypical structure of values described by Schwartz (1992).

is compatible with this social focus, but in contrast to Universalism, it has a narrower focus on preserving and enhancing the welfare of people close to oneself (family and close friends). Moving farther along the circle, Tradition (TR) values are also social in nature and focus on respecting, accepting, and committing to the customs and ideas of traditional culture and religion. Conformity (CO) refers to motivations of restraining actions or impulses that may upset or harm others and violate social expectations and norms. Next to Conformity, Security (SE) values emphasize concerns about safety, harmony and stability of society, social relationships, and the self. Therefore, these values start to mix both personal and social motivations. Power (PO) values capture goals of striving toward social status and prestige and controlling or dominating people and resources. Achievement (AC) emphasizes personal success through socially approved standards of competence. Moving more to the individually focused concerns, Hedonism (HE) values emphasize pleasure and sensual gratification of the senses. Stimulation (ST) captures motivations to seek excitement, novelty, and the pursuit of challenging goals in one's life. Completing the circle, Self-Direction (SD) values emphasize independent thought and action. These 10 values can be further differentiated. For example, Self-Direction values can be separated into a thought and action component, Security values have distinguishable social and personal security motivations, and Universalism values can be differentiated in terms of concerns about environmental and various social issues (Schwartz et al., 2012).

These 10 values can be clustered together in broader higher-order clusters along the two major dimensions. One major dimension places the value types of Self-Direction, Stimulation, and Hedonism into the Openness to change value cluster; the opposing cluster includes Security, Conformity, and Tradition (collectively labeled as conservation or conservative values). This clustering reflects a conflict between an overarching endorsement of one's own independent thoughts, actions, and interests versus an emphasis on restricting oneself and emphasizing the preservation of traditional practices and the status quo. This structure has some conceptual similarities with Hofstede's famous individualism-collectivism dimension.

The perpendicular dimension differentiates Self-Enhancement from Self-Transcendence values. Power and Achievement form the Self-Enhancement end of the value circle and are opposed to Benevolence and Universalism values at the Self-Transcendence end. Here the emphasis on pursuing personal success and interests (even at the expense of others), as well as motivations to dominate others and nature, is separated from a motivation to transcend selfish concerns and to promote the well-being of others and nature.

Schwartz (1992) argued that the overall value arrangement reflects two major motivational conflicts: (a) subordination to group pressure versus pursuit of own interests and (b) striving for success and social status versus a preference for equality and a concern for the welfare of others. Fontaine et al. (2008) hinted that different arrangements were also possible. For example, Benevolence, Conformity, and Tradition capture social orientations that are opposed to Stimulation, Hedonism, and Achievement, which emphasize personally oriented values. Universalism and Self-Direction emphasize values of personal and social growth, whereas Power and Security are concerned with personal protection and avoiding anxiety. Similar differentiations of the value space have been proposed by Gouveia and colleagues (2014).

The importance of this theoretical structure is that it enables theory-driven predictions about correlations of values with external variables (I discuss these predictions in more detail in the next chapter). Values are arranged in a two-dimensional space in relation to their assumed compatibilities and conflicts. Those that are motivationally compatible are located close to each other, and values that are conflicting (or unrelated to each other) are located at opposing ends.

Measuring Values in Psychology

In this section I briefly discuss how values are measured in psychology, demonstrating how this measurement is both compatible with and different from the measurement of traits. The Schwartz Value Survey (SVS; Schwartz, 1992) is the most widely used questionnaire. It captures the psychological importance of the value item for the individual in his or her subjective experience. It lists single-word value items or short phrases such as "freedom" or "devout" followed by further explanations of the term in brackets (e.g., "freedom of action and words" for "freedom"; "holding to religious faith & belief" for "devout"). These items are then rated on a 9-point scale, ranging from −1 (opposed to my values), 0 (not important), to 7 (of supreme importance). It is useful to keep the conceptualization and operationalization of a psychological construct aligned. Operationalized and measured as abstract life goals ("What is important in my life?") and considered in a cross-cultural context, the responses capture what is on people's minds and their subjective priority at that moment. In other words, measured values in the SVS provide an insight into the motivational goals that people deem important for themselves in a general sense.

The SVS is relatively abstract and context free, so Schwartz and colleagues developed the Portrait Value Questionnaire (PVQ; Schwartz et al., 2001, 2012). It differs from the SVS in that it describes prototypical individuals and asks respondents to indicate how similar this person is to him- or herself. Here is a typical item: "It is important to him/her to form his/her views independently." You may notice that this item is quite similar to a classic measure of personality traits. As discussed earlier, personality traits are often measured as typical behavior, and individuals are asked to agree or disagree with those descriptions. The major difference is that the PVQ attempts to measure the motivation underlying behaviors (it is important to behave in a particular way), and respondents have to decide whether this description describes them well or not. However, the main focus is still on the behaviors. With the introduction of this new tool to measure values, the differences between measuring values and personality traits have become quite subtle.

A Lexical Approach to Values

As indicated earlier, Schwartz's work was based on a limited set of value items. What would happen if we started off with a lexical approach to find common dimensions of values? There are three such studies in Dutch, Estonian, and German (Aavik & Allik, 2002; De Raad & Oudenhoven, 2008; Renner, 2003). In each language, researchers began by scanning dictionaries and then increasingly deleted items that did not refer to common definitions of values. They then removed items that were commonly considered to indicate ideologies or sexual orientation, and finally they removed synonyms. These lists of 666 Dutch, 560 Estonian, and 682 German words were subsequently rated by participants in terms of the importance of each term. The resulting data were factor analyzed (using the same methods as in Big Five studies), with the implicit assumption that there are independent value dimensions. Given the different starting points and methods, the overlap between these lexical value dimensions and the SVS is quite striking, as you can see in Table 3.3. You can also use this list to compare these values with the description of yourself or another person that you wrote earlier.

All major areas of the two main dimensions in the Schwartz circle are covered in these independent lexical approaches, even though not all value types described by Schwartz are included in each study. Some of the factor-analytic dimensions merged adjacent values from the Schwartz circle, leading to sometimes unexpected correlations. The religious and intellectual aspects as well as some of the social differentiations that are evident in the lexical studies are less well captured in the Schwartz dimensions. The correlations between lexical value factors and universal value lists employed by Schwartz varied, suggesting that there is scope for further refinement of the two-dimensional organization (see Schwartz et al., 2012). Yet, the general organization in his two major value dimensions seems to have been well replicated in these factor-analytical studies.

Value Measurement in Sociology and Political Science

As mentioned earlier, values have been studied in a number of different disciplines. In sociology and political science, the most prominent approach for capturing values of individuals (and cultures) has been the World Values Survey Project, initiated by Ronald Inglehart (Inglehart, 1971, 1990; Inglehart & Baker, 2000; Inglehart & Welzel, 2005). The initial value dimension in this line of research was the differentiation between materialism and postmaterialism values (Inglehart, 1997): it focused on the relative endorsement of goals emphasizing self-expression (input into government decisions) and quality of life (protecting freedom of speech) versus materialistic goals (fighting rising prices, ensuring the stability of the nation). Inglehart and Baker (2000) later refined the single materialism-postmaterialism dimension to create two major dimensions. The first dimension, Traditional versus Secular-Rational Authority, differentiated traditional moral and religiously based goals from more secular goals. Items included "How important is

TABLE 3.3 *Overlap between the SVS Dimensions and the Various Lexical Value Studies*

Schwartz (with the original value terms)	Renner: German value nouns	Renner: German value adjectives	De Raad & van Oudenhoven: Dutch values	Aavik & Allik: Estonian values
Power: Social power, wealth, authority, social recognition, preserving my public image, healthy		**Profit** (.42): rich, wealthy, achievement oriented, noble, materialistic, status conscious	**Status & Comfort** (.70): status, wealth, property, standing, beauty, reputation	
Achievement: Influential, intelligent, ambitious, successful, competent				**Self-enhancement** (.54): power, ambition, uniqueness, acknowledgment, respectability
Hedonism: Enjoying life, pleasure	**Profit** (.42): wealth, fortune, profit, being well-off, affluence		**Love & Happiness** (.72): love, spontaneity, cheerfulness, being sensitive, interpersonal warmth	
Stimulation: Exciting life, daring, varied life			**Competence** (.70): decisiveness, self-assuredness, enterprising spirit, change, adventure, exploration	**Hedonism** (.41): merrymaking, entertaining, adventurous, excitement, sexuality
Self-Direction: Independent, choosing own goals, curious, freedom, creativity, self-respect	**Intellectualism** (.29): social criticism, reflection, open-mindedness, freedom of education			**Self-realization** (.20): self-realization, independence, self-improvement, understanding, wisdom
Universalism: Wisdom, world of beauty, broad-minded, unity with nature, preserving the environment, true friendship, equality, inner harmony, social justice, world of peace		**Intellectualism** (.28): critical, progressive, political, cosmopolitan	**Aesthetics & Erudition** (.36); art, profundity, artistry, art of living, originality, creativity, music, culture	**Broad-mindedness** (.43): tolerance, creativity, spirituality, spontaneity, humanity

(continued)

TABLE 3.3 *(continued)*

Schwartz (with the original value terms)	Renner: German value nouns	Renner: German value adjectives	De Raad & van Oudenhoven: Dutch values	Aavik & Allik: Estonian values
Benevolence: Forgiving, honest, helping, responsible, loyal, meaningful life, spiritual life	**Balance** (.29): fairness, joie de vivre, human kindness, community, trust	**Balance** (.30): fair, warm, companionable, philanthropic, trusting		**Benevolence** (.43): complaisance, helpfulness, responsibility, sincerity, solidarity
Tradition: Humble, devout, moderate, accepting my portion in life, moderate, respect for tradition	**Salvation** (.45): blessing of god, faith in god, belief in god, firmness of faith	**Salvation** (.45): pious, blessed by god, devout, god-fearing, religious	**Benevolence** (.60): being accommodating, big-heartedness, being mild-mannered, charity, being good-natured, obligingness **Spirituality** (.32): religion, being religious, religious conviction, praying, god, morality	
Conformity: Politeness, respecting parents and elders, self-discipline, obedient	**Conservatism** (.55): National consciousness, national identity, love of one's country, sense of tradition		**Organization & achievement** (.59): conscientiousness, structure, organization, discipline, order, punctuality, thoroughness **Family & Tradition** (.26): marital duty, marital life, marriage, marital fidelity, parental duty, parental love, family ties	**Conservatism** (.38): rationality, industry, order, poise, neatness, thriftiness
Security: Family security, national security, sense of belonging, social order, clean, reciprocation of favors				

Note: Marker items and the highest positive correlation between the lexical factor and the Schwartz values are used to arrange the value dimensions; correlation is listed in parentheses.

God in your life?," "Showing greater respect for authority is a good thing," and "Abortion is never justifiable." The second dimension was labeled Survival versus Self-Expression Values. It captured some of the original ideas underlying the materialism versus postmaterialism index, focusing on the relative importance of personal freedom and quality of life versus concerns of physical and economic security. Items included a measure of general happiness and questions asking whether individuals have signed a petition, whether they see homosexuality as justifiable, and whether they think other people can be trusted. Chris Welzel (2013) extended and revised these dimensions, and his work will be of importance in Chapters 9 and 10, when I discuss value change. These value questions form the core of the World Value Survey, which has been administered to large representative samples worldwide. Every few years a new wave of data is collected, making this the largest data set on values covering the longest time period (starting in the mid-1980s).

Comparing Schwartz's and Inglehart's Dimensions

How similar or different are Ingelhart's two major dimensions from the dimensions described by Schwartz (1992)? Welzel (2010) developed a very short version of the PVQ that had been added to the World Value Survey, and its findings can be used to empirically examine the similarity of responses between Schartz's and Ingelhart's measurement approaches. Dobewall and Rudnev (2013) conducted this comparative analysis with data from the 2005 wave of the World Value Survey, which comprised data from 63,389 respondents in 46 national samples. I focus here on the key question of how the values of individuals are organized using the two different measurement approaches. One caveat that is important to note at the outset is that there are only single items available (which may not capture the full value spectrum) and the selection of individual items to represent Schwartz's theory might not always be ideal (for example, the very narrow focus on environmental protectionism when measuring Universalism values). With these issues in mind, let's look at Figure 3.2 that shows the plotted coordinates reported in Dobewall and Rudnev's analysis. As can be seen, all but one of the Traditional versus Secular-Rational authority items merged with the Openness to change values as measured by Schwartz. "Importance of God," one of the items from the Traditional versus Secular-Rational dimension emerged in the motivationally opposing Conservatism value cluster according to Schwartz. Therefore, the overall organization of

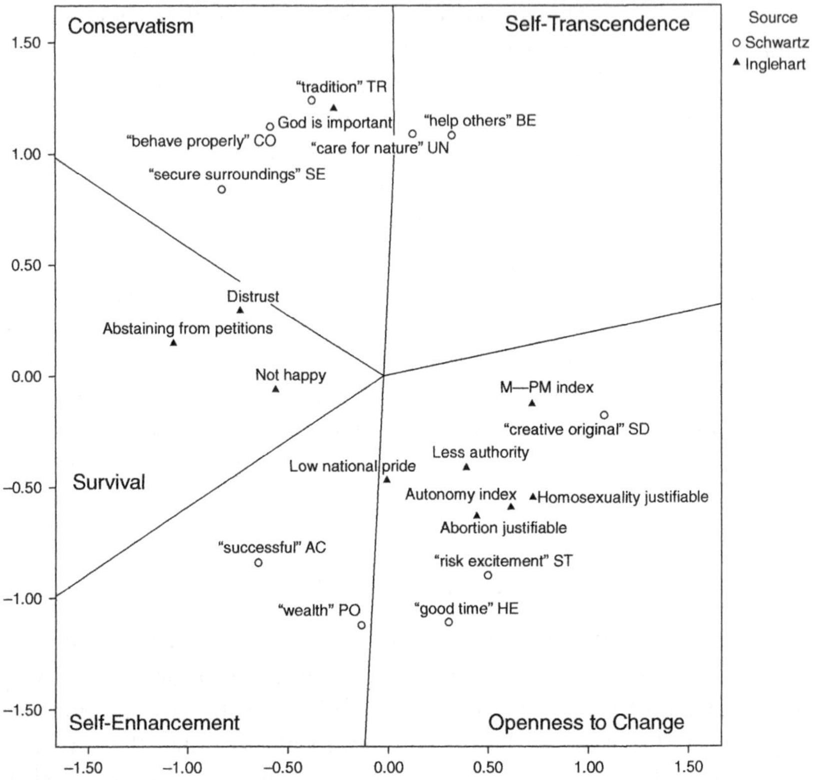

FIGURE 3.2 Overall similarity between the value dimensions proposed by Schwartz and Inglehart.

this dimension and the location of individual value items seem to be compatible between the measurement approaches.

Turning to the other dimension, two of the Survival versus Self-Expression items emerged with the Openness to change values from Schwartz. The remaining values from Inglehart's scale emerged closely related to Conservatism values and opposing the Openness to change items. Therefore, self-expression values as measured by Inglehart capture some conceptually similar content that is located very closely to the Openness to change cluster. Survival values in Inglehart's terminology emerged next to conservative values, highlighting some conceptual relation. Not engaging in political actions that may challenge the status quo (signing petitions) and being distrustful of others are compatible with conservative values as conceptualized by Schwartz. But these survival values nevertheless seem to capture some motivational content that is not captured by the 10 Schwartz-type items that were included in this study. Happiness in particular has no conceptual equivalent in Schwartz's research. Consequently, Self-Enhancement versus Self-Transcendence seems to be only clearly measured with items derived from the Schwartz theory. The two dimensions used in the Inglehart scale seem to capture finer distinctions in Schwartz's Openness to change versus Conservatism value dimension.

One interesting avenue for future research is to examine these dimensions' relative similarity and overlap more carefully, using a more extended set of values. What I can conclude cautiously is that, despite very different approaches to measuring values and even with very limited measures of values, we nevertheless still find a spatial organization that captures the main two motivational orientations that I described earlier. The value research in the political domain can help us make finer distinctions between motivations that underlie the Openness to change versus Conservatism axis, a dimension that is of great interest to political scientists. The Self-Enhancement versus Self-Transcendence dimension in contrast is driven by values that were derived from Schwartz's work.

SUMMARY

Traits and values describe differences between individuals. Personality traits refer to what people are like (their behavioral tendencies or reaction norms), whereas values describe what people consider important (e.g., their conscious goals). Research on traits and on values both arose from the lexical hypothesis, which proposes that major dimensions of personality can be captured in single words (or word clusters) in most languages. In line with this common origin, both personality and values are systematically organized. Five major personality dimensions – Extraversion, Neuroticism (or Emotional Stability), Conscientiousness, Agreeableness, and Openness – have been identified across a large number of studies. Values have been organized in a two-dimensional space around compatibilities and conflicts between major underlying goals. Most often, 10 value types are differentiated that are organized into two higher-order dimensions: Openness to change versus Conservatism values and Self-Enhancement versus Self-Transcendence values. As mentioned earlier, you can use Tables 3.1 to 3.3 to compare your description of a person with the major dimensions of personality and values identified in these lines of research. You could also go back to my description of Tina and Sandra in Chapter 1 and think about what words you could use to describe their behavior.

When comparing these tables, you may have noticed some conceptual and empirical overlap. Up to now, I have discussed these two approaches for describing individuals' traits and values separately. In the next chapter, I turn to the extent to which these two approaches are compatible with each other.

4

Examining the Common Structure of Traits and Values

How similar are values and personality traits? Do individuals who endorse a certain value, such as Benevolence, as important in life, show distinct personality traits? Do extroverts or agreeable individuals value different things in life compared to introverts or dominating individuals? Personality traits in the Big Five tradition are the reaction norms or typical behaviors across a range of situations, whereas values are conceptualized as motivational goals – what people find important in their lives independent of the situation. In this chapter, I present evidence that these two approaches are actually quite compatible with each other. First, I describe studies that have examined the correlations between the 10 value types described in the previous chapter and personality traits in the Big Five tradition. Noting the relatively consistent correlations between values and personality, I then revisit the independence of trait assumption in the personality literature. Here, I draw on recent discussions of the so-called meta-traits, which are a second-order clustering of the five traits (DeYoung, 2006; DeYoung, Peterson, & Higgins, 2002; Digman, 1997). I report reanalyses of trait correlations that show a systematic spatial organization along two major axes. Bringing the value–personality correlations back together with the idea of meta-traits and their spatial organization, my argument is that there is common structure to both values and personality traits, which points to some common underlying motivational core of values and traits. But this is jumping ahead to the next chapter; let's begin by examining the relationship between values and traits. If you like two-dimensional figures, this chapter will make you happy.

HOW SIMILAR ARE VALUES AND BIG FIVE TRAITS? AN EMPIRICAL APPROACH

Together with Diana Boer, I conducted a meta-analysis of studies that had reported correlations between values and FFM-relevant personality traits (Fischer & Boer, 2015). A meta-analysis is a quantitative summary of all available research that provides a comprehensive and rigorous overview of the state of the science. We searched for published and unpublished research that used either version of the value surveys that Shalom Schwartz had developed (which I discussed in detail in Chapter 3) and correlated value scores with any version of the Big Five personality trait inventories. We suspected that there would be similarity between the Big Five personality traits and the 10 value types, and we wanted to substantiate that. As shown in the previous chapter, both approaches can be traced back to the lexical hypothesis. Furthermore, there seems to be much conceptual and theoretical overlap between some of the traits and values. This overlap is most obvious with Openness and Openness to change, where the trait and value terms are nearly synonymous. Science can only progress if we dispassionately examine the relationship between constructs to see how they are related and so remove redundancies. This effort also supports the principle of parsimony, finding the simplest explanation for empirical relationships.

There are more benefits to examining the relationships between traits and values. As I discussed in the previous chapter, values are organized in a two-dimensional space based on their compatibilities and conflicts. Because of this structure, relationships between any value type and third variables show systematically increasing and decreasing correlations as we move around the value space. This circular organization can help identify underlying patterns in data, even if the original researchers may not have examined these possibilities. Sipke Huismans suggested this structure to Schwartz (see footnote 16 in Schwartz, 1992); it provides a powerful way to examine the underlying structure of psychological attributes. If researchers fail to pay attention to this circular organization and instead examine each correlation in isolation, they may come up with very different conclusions based on the same evidence (Parks-Leduc, Feldman, & Bardi, 2014).

Advantages (and Desirable Properties) of the Circular Value Structure

To explore the advantages of the circular organization of data, Diana Boer and I (Boer & Fischer, 2013; Fischer & Boer, 2015) studied the statistical properties of psychological systems organized in a circular structure. There

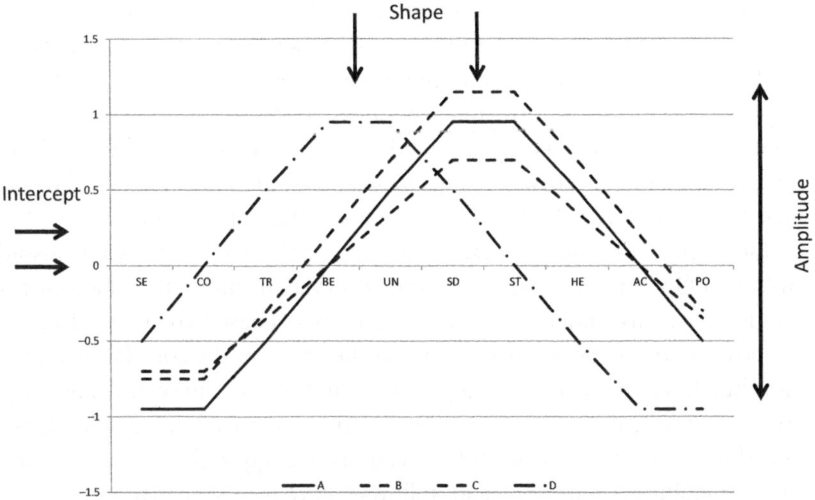

FIGURE 4.1 Examples of different correlational profiles.

are three important parameters of the correlations between values and other constructs: amplitude, shape, and intercept. Figure 4.1 depicts those three parameters.

Amplitude is the difference between the peak and the trough of the pattern, corresponding to the maximally positive and negative correlations. It thus captures the strongest empirical relationship between any objects under study, in this case values and traits. If there are both strong positive and negative correlations between values and traits, then this particular trait is strongly related to the motivational forces that underlie the value structure. The second parameter is the shape, which is the relative positioning or patterning of the peak and trough of the correlational profile. For example, because there are 10 value types, the peak and trough of each one's correlation with traits might be positioned at different points along the circle. If the peak is positioned at Benevolence, this trait obviously has a stronger association with motivations to help others close to them than if the peak is positioned at Self-Direction. The final parameter is the intercept or baseline, which is simply the average correlation. It captures how much all the correlations are affected by some variables that are unrelated to both values and traits. A prototypical case is the presence of method bias (all responses are based on self-reports, which may shift the responses independently of the content of what is being asked). Method bias might be due to general response tendencies; for example, giving a more positive answer to any question just to put oneself in a more favorable light or because of carelessness. These shifts may not be bias and some shifts can also be meaningful. One example could be some general motivational differences between respondents (e.g., some individuals might be more attuned to their personal values or more motivated by any kind of goal compared to others who do not care much about their personal values). For these individuals, the correlation between all values and all traits would be much stronger.

Table 4.1. shows four curves. Curves A and D show the typical sinusoid curves for Openness to change versus Conservatism, and Self-Enhancement versus Self-Transcendence correlation patterns. These two curves differ in the position of the peak correlations (in the shape of the correlational profile), but have the same intercepts and amplitudes. Curve B follows the same prototypical shape as curve A, but has a reduced amplitude; therefore, this particular trait does not correlate as strongly with values (assuming equal reliabilities). Curve C again follows Curve A in shape (therefore captures the same motivational content), but all correlations are shifted upward with slightly stronger positive correlations and slightly weaker negative correlations.

The analysis of these correlational profiles can tell us (a) how systematically traits and values are related, (b) how strong this association is, and (c) if there are unmeasured third variables or method problems that might influence any correlation. Adopting this theory-based approach thus enables us to understand more fully how other constructs such as personality are related to values. Even small or close to zero correlations become meaningful if they are embedded within a systematic organizing pattern. A systematic pattern with relatively strong positive and negative maximal correlations indicates that values and traits are likely to be closely related and organized by similar underlying mechanisms. Studying these correlational profiles can give us valuable insights into the structure of personality more generally.

Finding the Relevant Evidence

For our meta-analysis of values and personality, we focused on the relative position of the peak and trough; that is, the shape of the correlation pattern. For this reason, we only included studies that had measured all 10 value types and their associations to a Big Five personality trait; unfortunately, to date no studies have examined the relationship between the 10 trait aspects described by DeYoung and colleagues (DeYoung et al., 2007) and values. We also only included studies with nonclinical samples of

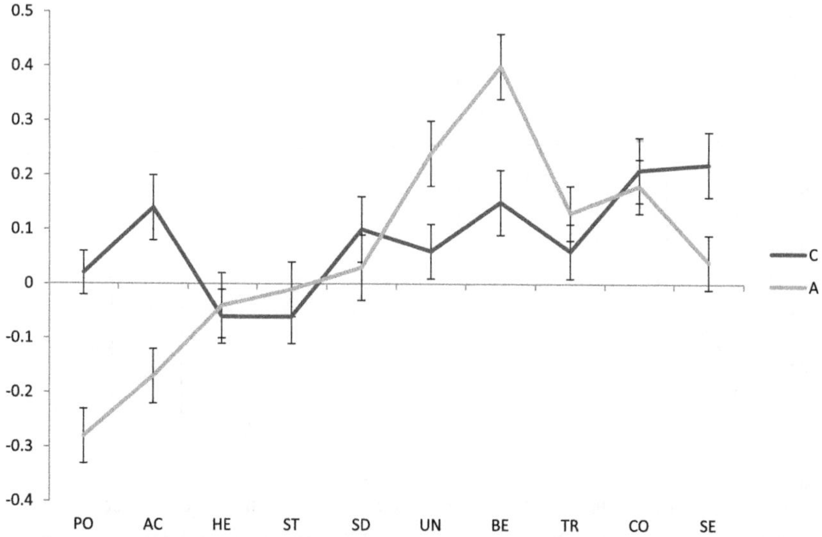

FIGURE 4.2 Correlation between values and Conscientiousness and Agreeableness traits. *Notes:* C = Conscientiousness, A = Agreeableness, PO = Power, AC = Achievement, HE = Hedonism, ST = Stimulation, SD = Self-Direction, UN=Universalism, BE = Benevolence, TR = Tradition, CO = Conformity, SE = Security.

participants over the age of 16. As is often the case, not all studies reported sufficient data to compute effect sizes, and not all authors whom we contacted to provide this information did so. Despite these constraints, we were able to include 26 samples with 9,935 participants from 14 countries. Most of these studies (58%) used students as participants, as is common in psychological research. Every reported correlation coefficient in previous research was weighted by sample size so that bigger samples had a larger impact. After weighting all correlations by the sample size (because larger samples are more accurate and should therefore have a bigger influence on our results), we plotted the average correlations between values and personality traits, as shown in Figures 4.2 to 4.4.

Correlation Profiles of Values and Big Five Traits

I discuss each personality trait in turn. The Agreeableness trait was positively correlated with Self-Transcendence values (Benevolence, Universalism) and negatively with Self-Enhancement values (Power, Achievement). The strongest correlations were between Agreeableness and prosocial

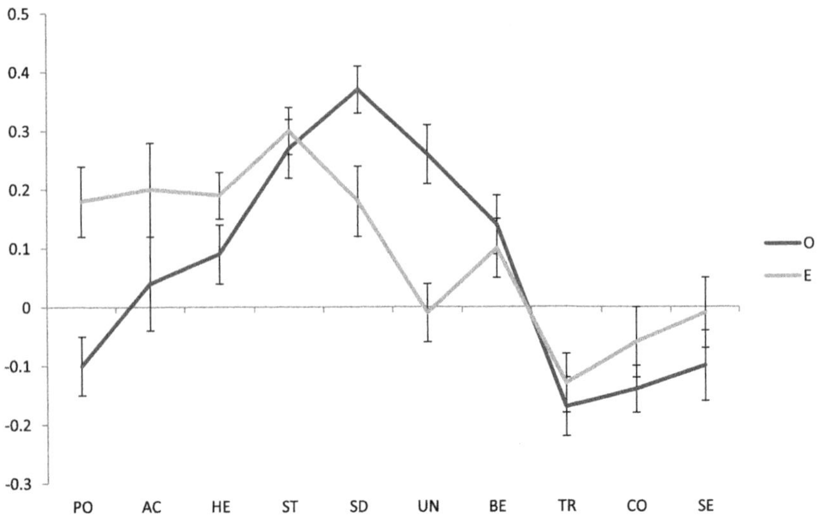

FIGURE 4.3 Correlation between values and Extraversion and Openness traits. *Notes:* C = Conscientiousness, A = Agreeableness, PO = Power, AC = Achievement, HE = Hedonism, ST = Stimulation, SD = Self-Direction, UN=Universalism, BE = Benevolence, TR = Tradition, CO = Conformity, SE = Security.

in-group focused values (Benevolence, positively) and self-centered status and power-oriented values (Power, negatively). This indicates clear and consistent associations of Agreeableness with Self-Transcendence versus Self-Enhancement values. We found no consistent associations with values along the Openness to change versus Conservatism axis.

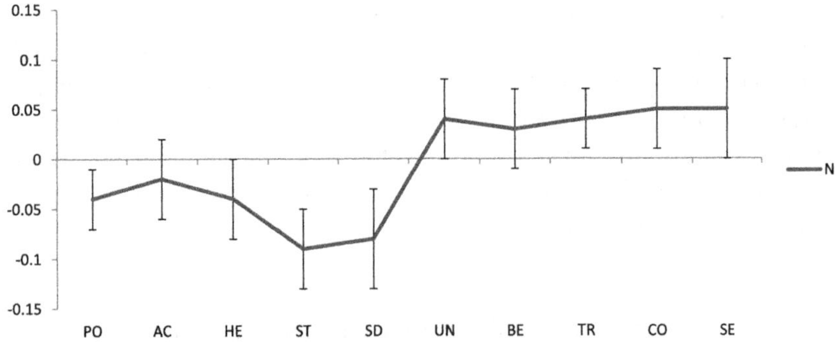

FIGURE 4.4 Correlation between values and Neuroticism. *Notes:* C = Conscientiousness, A = Agreeableness, PO = Power, AC = Achievement, HE = Hedonism, ST = Stimulation, SD = Self-Direction, UN=Universalism, BE = Benevolence, TR = Tradition, CO = Conformity, SE = Security.

Conscientiousness was positively associated with Achievement values, but the most consistent associations were positive correlations with conservative values (primarily with conformity and Security) and negative correlations with Openness to change values (Hedonism, Stimulation). Therefore, Conscientiousness is relatively consistently aligned with values along the Openness to change versus Conservatism dimension, but also shows some secondary positive associations with values related to achievement striving and status. This suggests that Conscientiousness is related to both major value dimensions, with a slightly stronger association to the Openness to change versus Conservatism axis.

Moving to Figure 4.3, the Openness to experience trait was positively correlated with Openness to change values (mainly Self-Direction and Stimulation, but also Hedonism) and negatively with conservative values (Conformity, Tradition, but also Security). The strongest positive and negatives correlations between the Openness to experience trait were with Self-Direction and Tradition values, respectively. Hence, Openness as a personality trait shared some strong communalities with values capturing motivations to seek out independent thoughts and actions, as well as the rejection of traditional rules and norms of society. The value profile was very strongly aligned with the Openness to change versus Conservatism value axis. There were also some positive correlations between this trait and Universalism and Benevolence values. The pattern of alignment between Openness as a trait and prosocial concerns along the Self-Transcendence versus Self-Enhancement value axis was not highly consistent and not practically meaningful; it did not reach the threshold set by Boer and Fischer (2013). Therefore, Openness traits showed a clear association only with the Openness to Change versus Conservatism dimension.

Extraversion showed the strongest correlations with Stimulation (positively) and Tradition values (negatively). The overall profile was clearly correlated with the Openness to change versus Conservatism values axis, and this alignment was stronger than that for Conscientiousness discussed earlier. There were also some moderately strong and positive correlations with the self-enhancing Achievement and Power values. However, there was no overall alignment of the correlation patterns with the prototypical Self-Enhancement versus Self-Transcendence profile. Therefore, some residual motivations might be associated with Power and Achievement values, but they were neither clearly nor systematically aligned with this second value axis. Therefore, the overall strongest association was with Openness to Change versus Conservatism values.

Last but not least, Neuroticism was only weakly associated with value orientations: there were small negative correlations with Openness to change values (Hedonism, Stimulation, Self-Direction) and positive associations with Conservatism values (Security, Conformity, Tradition). In terms of the two major value dimensions, a significant association with Openness to change versus Conservatism values emerged, but did not exceed what we had deemed practically meaningful in the interpretation of the profiles (Boer & Fischer, 2013). These associations were therefore not systematic, yet they followed the overall pattern of compatibilities and conflicts inherent in the value circle.

Recapping, the motivationally clearest and most consistent associations were between Agreeableness and self-transcending values (especially Benevolence) and Openness and Openness to change values (in particular, Self-Direction and Stimulation). Conscientiousness and Extraversion also showed clear alignments with the Openness to change to Conservatism value axis and some secondary correlations with Achievement and Power values. Therefore, both traits were located on opposing ends of the primary value axis, but shared some motivational content with Achievement and Power values (that are located on the second underlying value axis). The weakest associations between personality traits and values were found for Neuroticism, but even these associations still suggested a systematic and nonrandom pattern.

Shifting the attention to the amplitude parameter, the strongest overall correlations were found for Agreeableness, followed by Openness to experience, Extraversion, Conscientiousness, and finally Neuroticism. To put these correlations into context, we need to consider the correlations between specific personality instruments that measure conceptually overlapping traits (e.g., how strongly correlated are Agreeableness traits when measured with the NEO-PI-R versus the BFI) and compare these associations with the strength of correlations between trait and values. Pace and Brannick (2010) reported the empirical overlap between personality traits measured with different instruments, and we can use these meta-analysis as a benchmark for personality trait measurement. As discussed in the previous study, there are fewer value instruments available, and to date, there is no comprehensive study to assess the relative similarity of value instruments across a larger number of studies. However, there have been studies of correlations between value measurements in single studies. When developing his alternative measure of the PVQ (see Chapter 3), Schwartz and colleagues reported correlations varying between .44 (Conformity) and .70 (Stimulation). Lindeman and Verkasalo (2005) in a study of Finnish youth found

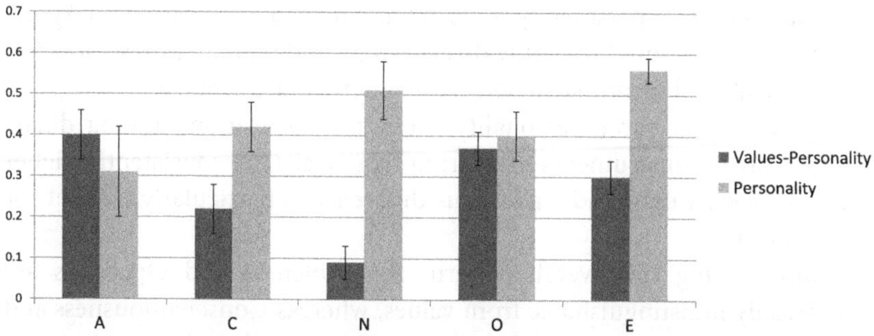

FIGURE 4.5 Average value correlations of different personality inventories (Pace & Brannick, 2010) compared to highest average value-trait correlation (Fischer & Boer, 2015). *Notes:* A = Agreeableness, C = Conscientiousness, N = Neuroticism, O = Openness, E = Extraversion.

correlations varying between .52 (conformity) and .78 (Universalism). Yet, when using lexical value scales as comparison (see Table 3.3), the correlations between conceptually related values could be as low as .20. Therefore, these correlations between different instruments measuring conceptually overlapping values and trait content set the upper boundaries for what we might find when correlating values and traits with each other. With these boundaries in mind, let's have a look at the relative strength of the value-trait correlations.

Figure 4.5 contrasts the typical correlations between personality tests measuring the same conceptual Big Five personality trait across different instruments[1] (Pace & Brannick, 2010), with the highest average correlation (using the absolute correlation coefficient value) between values and personality traits found in our meta-analysis. What is striking to see is that the correlations between values and Agreeableness as a personality trait that we found are actually slightly higher than between different personality tests measuring Agreeableness in general. For Openness to experience, the correlations are highly similar. In both cases, the standard errors overlapped, suggesting that the correlations are not statistically distinguishable. There are two possible explanations for this pattern. The first is that both Agreeableness and Openness to experience traits are very similar to value orientations and therefore share a strong empirical overlap. The second plausible

[1] I used the correlation estimates for all personality inventories, instead of using just correlations based on instruments that explicitly measured the five-factor framework, because in our meta-analysis of value-personality correlations we included other personality inventories. We recoded these personality dimensions to match the definition of the Big Five.

explanation is that personality traits of Openness to experience and Agreeableness are measured in highly diverse ways, and therefore they show relatively weak and inconsistent correlations across instruments.

For the other three personality traits, the correlations among different personality instruments measuring these traits are consistently higher than between values and traits. This difference is particularly evident for Neuroticism.

Summarizing the overall pattern, Agreeableness and Openness are empirically indistinguishable from values, whereas Conscientiousness and Extraversion are systematically related with values, but these traits measure additional content that is not shared by values. Neuroticism is also weakly but systematically associated with values, but it appears to capture personality dynamics that are somewhat distinct from the motivational concerns that drive values. If you did the exercise presented in the beginning of Chapter 3, you may have found, for example, that you listed words that indicate that you are high in Agreeableness and also that you value Benevolence and Universalism-type values.

Values and HEXACO Trait Correlations

The previous section focused on correlations between values and the Big Five traits. As you may recall from Chapter 3, the HEXACO model conceptualizes the five factors somewhat differently (redistributing the content between Neuroticism and Agreeableness) and adds a sixth factor, which measures Honesty-Humility. How similar or different are values and traits if we consider this alternative interpretation of the trait structure? Lee, Ashton, and colleagues reported data from two large samples in North America (Lee et al., 2009). Their analyses used slightly different methods and the correlation matrix was not reported in the paper, so we cannot directly compare the correlations in the same way that our meta-analysis did. However, from their analysis we can infer that Honesty-Humility related very consistently to the Self-Transcendence versus Self-Enhancement dimension in the value space. Honesty-Humility traits most strongly differentiated between Power (negatively) and Benevolence values (positively). Openness to experience in contrast showed strong overlap with the Openness to change versus Conservatism value dimension, with the strongest contrast being between Self-Direction and Conformity. Lee et al. interpreted this similarity as indicating that these two personality traits – Honesty-Humility and Openness to experience – are closely related to desirable goals; that is, they are conceptually similar to values. And vice versa, the way these values that highly correlate with traits are typically measured (in particular, Self-Direction) makes them highly similar to these personality traits. Therefore, using this alternative interpretation of traits, values and traits also show strong empirical relationships.

REVISITING THE TRAIT STRUCTURE: THE BIG TWO

In the previous chapter, I discussed the factorial dimensions of the Big Five. Traditionally, personality traits have been assumed to be independent. At the same time, these supposedly independent dimensions showed some meaningful correlations with values that follow a circular structure. This requires us to revisit the independence assumption of the five factors. In fact, this independence assumption has been questioned in a number of studies over the years. Digman (1997) was the first to draw substantive attention to a possible higher-order structure. Reanalyzing data from 14 different studies, he reported that two higher-order dimensions consistently emerged in a second-order factor analysis. He called the first factor Alpha, which had high loadings from Agreeableness, Conscientiousness, and Neuroticism (negatively). The second factor was labeled Beta and consisted of Extraversion and Openness to experience. Digman speculated that the first factor might be a socialization factor, whereas the second factor could be interpreted as a personal growth factor. DeYoung and colleagues (DeYoung, 2006; DeYoung et al., 2002) replicated these two meta-dimensions, but gave them different labels. Because Alpha in their view was concerned with stability in the emotional domain (Neuroticism), social domain (Agreeableness), and motivational domain (Conscientiousness), they dubbed it Stability. Beta was interpreted as a Plasticity factor because it referred to behavioral (Extraversion) and cognitive (Openness) flexibility. DeYoung and colleagues also proposed some specific neurobiological mechanisms that led to these higher-order factors, which I discuss in more detail in the next chapter.

Combining these two meta-traits with the trait aspects that I discussed in the previous chapter, we can see a hierarchy of traits. At the highest level, there are the two meta-traits, whether labeled Alpha or Beta, or Stability or Plasticity. Below them are the traditional Big Five, which in turn separate into the 10 trait aspects. Below the trait aspects are an unknown number of trait facets that are directly measured by questionnaire items. In Figure 4.6, I display this hierarchy from the inside out, with meta-traits in the center,

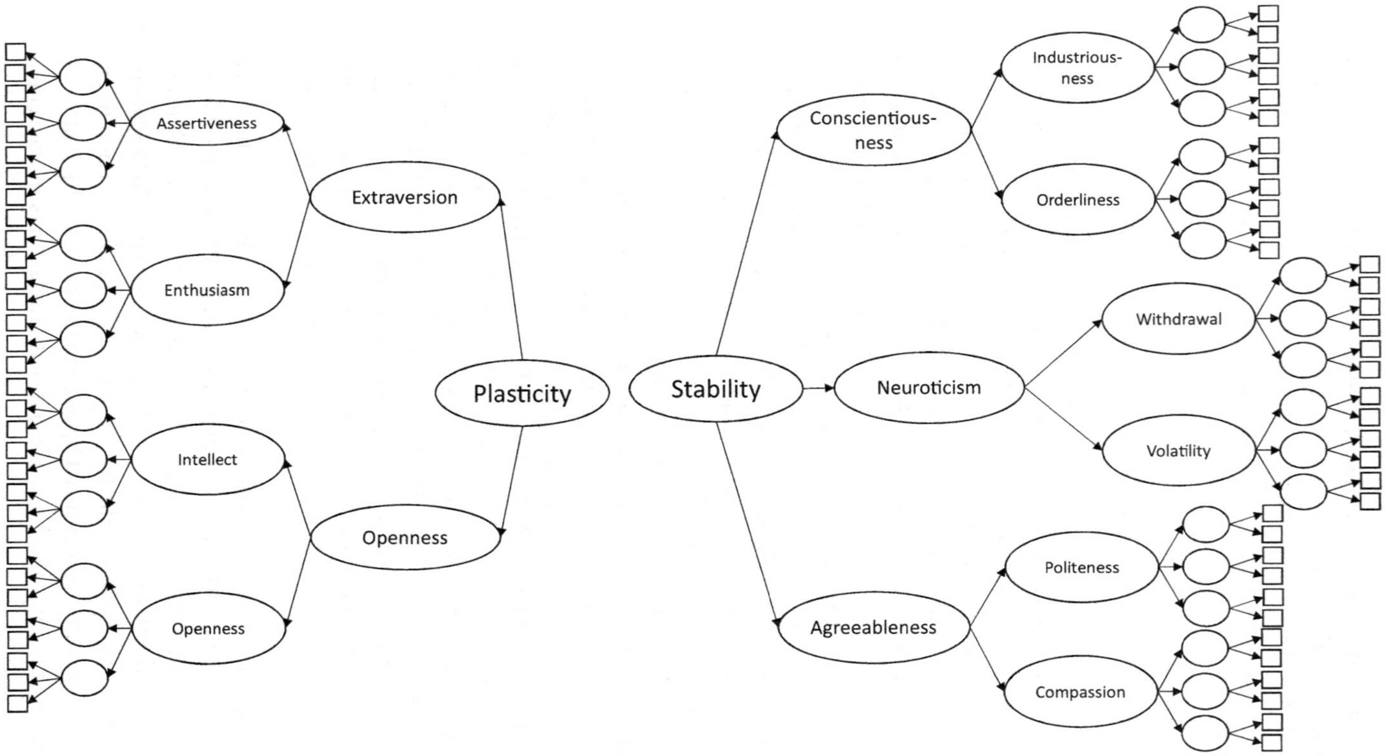

FIGURE 4.6 The hierarchical structure of traits. *Notes:* The two meta-traits (Stability and Plasticity) are positioned in the center, forming the apex of the personality structure. Traits form the second layer, integrating trait aspects (two per trait), which in turn sit on top of trait facets (unknown number), which are measured by individual items (small boxes). I use boxes for personality items, because they can be directly observed (responses by participants in personality studies) and ovals for the assumed latent variables (that are inferred from responses to questionnaire items).

followed by traits, their constituting trait aspects, and then the larger number of trait facets, which each can be measured by a larger number of items (which are placed at the outside).

A DIFFERENT LOOK AT PERSONALITY STRUCTURE: USING MULTIDIMENSIONAL SCALING TO UNCOVER UNDERLYING PERSONALITY STRUCTURES

Working around the same time but independently of DeYoung's, Digman's, and their colleagues' work, Maraun (1997) used multidimensional techniques similar to the ones used by Schwartz (1992) to examine the underlying structure of the NEO-PI-R. Using the correlation matrix provided in the NEO manual (Costa & McCrae, 1992), he found that two dimensions are sufficient to describe the relations between the supposedly independent five dimensions. Moving around the two-dimensional space, Conscientiousness was located adjacent to Extraversion, followed by Openness, Neuroticism, and Agreeableness. Neuroticism was most distant from the center of the two-dimensional space, suggesting that it is most different from the other traits. Most strongly opposing Neuroticism was Conscientiousness, suggesting a stability versus instability dimension. Maraun replicated this general two-dimensional structure using trait adjectives (Goldberg, 1990), with only Extraversion and Openness swapping places.

Can we replicate this circular organization of traits with different Big Five measures? As mentioned, De Young and colleagues (2007) had described and measured 10 personality aspects that are situated below the Big Five in the personality hierarchy (see Figure 4.7). They administered their new scale together with an established Big Five inventory (Benet-Martínez & John, 1998; John, Naumann, & Soto, 2008) to student and community samples in North America; this provided an interesting opportunity to examine the structural relations in two independent samples.[2] Figure 4.7 shows the spatial organization of these traits and aspects. The patterns across these two sample clearly replicated the ordering of the NEO-PI-R arrangement reported by Maraun (1997). Moving around the two-dimensional space,

[2] For those interested in technical details, I analyzed these correlation matrices as similarity matrices for multidimensional scaling. I used proxscal with a Torgerson initial configuration and assumed interval-level data. Minimum stress and stress convergence were set to .0001 with a maximum of 100 iterations. Because there are two different data sets and each analysis might show arbitrary sampling-driven differences (e.g., arbitrary rotations, shrinkage, dilution, or inversion of the major axes), I rotated the results of each analysis so that the dimensions are directly comparable.

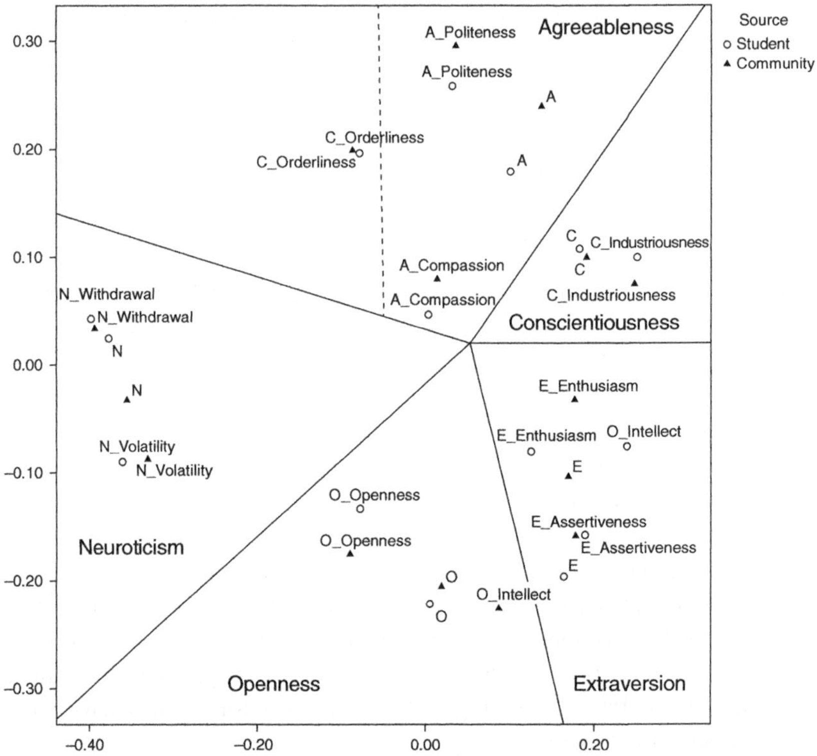

FIGURE 4.7 Spatial organization of Big Five traits and aspects. *Note:* Single letters (O, C, E, A, N) refer to the Big Five means (John, Naumann, & Soto, 2008), and single letters linked by underscore to a trait aspect name (e.g., A_Politeness) refer to the Big Five Aspect means (DeYoung et al., 2007). See Table 3.2 for a description of these trait aspects.

Agreeableness is located next to Conscientiousness, then Extraversion, and Openness, and the circle closes with Neuroticism. The only unexpected location was for the trait aspect "orderliness," which is part of Conscientiousness: it was located between Agreeableness and Neuroticism, suggesting that orderliness may be motivationally closer to Neuroticism. This may make sense if we consider that orderliness can help individuals with higher Neuroticism scores to calm down and control their emotions (see also that the withdrawal aspect of Neuroticism is positioned close to orderliness, which underscores this conceptual relationship). Among students, intellect as an aspect of Openness was associated with enthusiasm, which is an Extraversion aspect. This is an interesting location, because it suggests that the social role of students may somewhat shift the organization of their personality traits. I discuss the extent to which environmental factors influence the organization of values and traits in much greater detail in Chapters 7 and 8.

The location of the trait aspects also reveals further interesting information. For example, orderliness and politeness are located close to each other in this two-dimensional space. Being polite is also intuitively associated with order, because being polite helps maintain order in social relations. This close location suggests that politeness and orderliness may tap into some common motivational core.

Can we identify similar structures using the HEXACO inventory? As with the Big Five aspects, it is also possible to distinguish lower-level aspects in the HEXACO model (Ashton, Lee, Goldberg, & de Vries, 2009). Having more information available on these aspects is an advantage, because it may allow us to examine the spatial organization of traits in finer detail. Ashton and colleagues (2009) reported correlation matrices of the HEXACO trait aspects model separately for Canadian, US, and Dutch samples. Figure 4.8 shows the results of the same MDS procedure that I described earlier. As before, relatively clear clusters emerged. Honesty-Humility and Agreeableness were mingled, suggesting that these two components might be much more similar when examined with a technique that explicitly allows relationships between factors. I noted their conceptual similarity in Chapter 3. Conscientiousness and Emotionality subscales were also located very close to each other. This might not be surprising because Emotionality was positively phrased, and in the previous analyses Conscientiousness and Neuroticism (as a negatively valenced trait) were located in opposition to each other. There was a clear differentiation of Openness and Extraversion on the positive pole of dimension 1 from Agreeableness/Honesty, Emotionality and Conscientiousness on the negative pole of dimension 1. Similarly, Agreeableness was located diagonally opposite to Extraversion along dimension 2. Consequently, the spatial organization of the HEXACO model again points toward a systematic underlying organization of traits, similar to values. These independent studies using different instruments and samples all suggest a two-dimensional underlying structure of personality traits. Extraversion and Openness traits were typically located next to each other, resembling the Plasticity meta-trait. Conscientiousness and Agreeableness were also often (but not always) located in proximity, with Neuroticism (versus emotional stability) emerging either in opposition to these two traits (if phrased as negatively valenced Neuroticism) or in the vicinity of these two traits (if phrased as emotional stability).

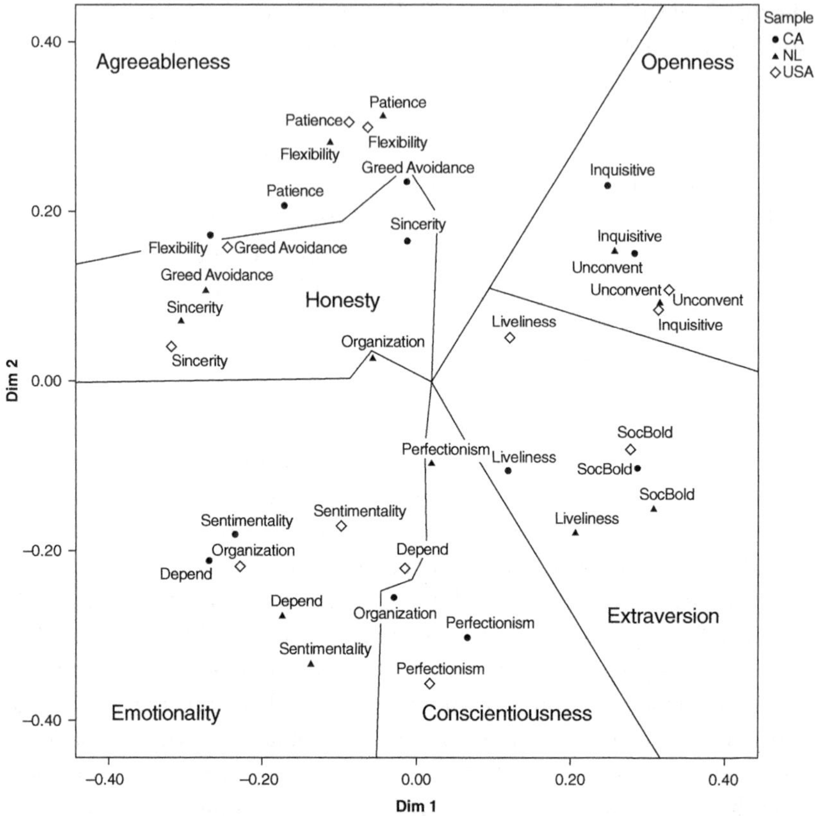

FIGURE 4.8 Spatial organization of the HEXACO traits and aspects. *Notes:* Depend = Dependence, SocBold = Social boldness, Unconvent = Unconventionality, Inquisitive = Inquisitiveness.

REINTEGRATING THE VALUE-PERSONALITY CORRELATIONS WITH THE TRAIT STRUCTURE

You may have already spotted some overlap between the two-dimensional organizations of values and the spatial organization of traits. On one side is a cluster of personality scales measuring Openness to experience, and Extraversion. Agreeableness, Conscientiousness, and emotional stability (versus Neuroticism) tend to emerge in a separate cluster. Depending on how the specific trait is measured in a specific instrument, the two meta-traits are organized in slightly different constellations. For example, if Neuroticism is phrased positively as Emotionality or Emotional Stability, it will emerge close to Conscientiousness. However, this basic distinction seems to emerge consistently across a large number of studies (Strus, Cieciuch, & Rowiński, 2014b). In addition, ignoring Neuroticism for the time being, the relative ordering across different instruments moving around the outer edges of this two-dimensional space is from Conscientiousness to Agreeableness, Openness, and then Extraversion. This ordering is relatively matched with the highest correlating values, moving from conservative values to self-transcendence, Openness to change, and Self-Enhancement.

When examining the specific value-personality correlations, the plasticity personality traits (Openness and Extraversion) correlated slightly more strongly with values overall than did the stability traits (in particular Conscientiousness and Neuroticism). The two plasticity personality traits are relatively neatly aligned along the Openness to change versus Conservatism value axis. Extraversion also shows some secondary correlations with Self-Enhancement.

As mentioned, stability meta-traits overall were somewhat less consistently aligned with values. Agreeableness is most strongly aligned with the Self-Transcendence versus Self-Enhancement value dimension. Conscientiousness is a stability meta-trait that is motivation oriented (DeYoung et al., 2002), but is more closely associated with values along the Openness to change versus Conservatism axis (Fischer & Boer, 2015). As reviewed earlier, the strongest correlations between Conscientiousness and values are observed for Security and Conformity values from the Conservatism cluster. These values are concerned with maintaining the status quo, providing security for one's loved ones and important social groups, and not upsetting the social and moral order. Therefore, it is the avoidance element that drives these correlations. Finally, Neuroticism is the personality trait that correlated only weakly with values, but there is nevertheless a systematic pattern that suggests a motivational alignment with Openness to change versus Conservatism values. Here, avoiding anxiety by not taking part in stimulating, novel, and risky activities and engaging in conformity-oriented behavior might be the driving force behind these correlations.

I considered the value-trait correlations in light of the correlations that we find when correlating traits or values measured with different instruments with each other. Even if the theoretical content is the same, each instrument measures it slightly differently and it is unlikely that we find perfect correlations in the real world. These measurement dependencies of correlations impose an upper limit of what we might find when correlating values and traits directly with each other. We give participants quite different instructions when rating values versus personality traits – tell us what is important for you in your life versus describe how you typically behave.

Given these very different instructions and ways of measuring values and traits, the empirical similarity shown in this chapter is striking.

SUMMARY

The closest associations between values and traits are found for plasticity meta-traits. The plasticity meta-traits align quite nicely with Openness to change versus Conservatism values. Agreeableness (as well as Honesty–Humility) as a socially focused stability trait is closely related to values along the Self-Transcendence versus Self-Enhancement value axis. The specific motivational focus of the remaining traits appears to be important. Extraversion is located closer to the Openness to change pole, whereas Conscientiousness is positioned toward the Conservatism end of the same dimension. Both traits seem to share some secondary motivational content with Achievement and Power values, but there is no overall systematic relationship with the underlying dimension.

There is considerable empirical overlap of personality traits and values (Strus, Cieciuch, & Rowiński, 2014b): certain traits and values seem to go together. You may have noticed this when reviewing how you described yourself or another person in the exercise at the beginning of Chapter 3. Returning to Tina and Sandra whom we met in Chapter 1, they displayed some behaviors that seem to suggest correlated traits and values. For example, Sandra was closely attached to her family and showed signs of being highly conscientious, clean, and tidy (and potentially a wee bit neurotic, considering how carefully she washes her cup and places it back into a locked cupboard on her desk). You may wonder (as I do) what might be the reasons for these associations between both values and traits and also among the different traits. Are there some possible underlying mechanisms that can explain these associations? In the next chapter, I begin to address this question. What are possible mechanisms that can explain the similarity between values and traits?

5

Explaining Personality Structures: The Relative Importance of Genetic and Cultural Differences in Values and Traits

In the previous chapter, I demonstrated that traits and values are correlated and that both traits and values are systematically organized in a shared two-dimensional space. How can we interpret such patterns? In this chapter, I make my first attempt to bring together different theoretical approaches that can help explain both the structure of personality and the relationship between values and personality traits. These various accounts refer to both genetic and cultural variation. I present data from twin studies and cross-cultural comparisons to estimate the relative variance due to genetic and cultural processes. At the end of the chapter, we have sufficient insights to make a first guess about how similar or different are Tina's and Sandra's personality traits and values.

EXPLAINING PERSONALITY: A HIERARCHY OF PERSONALITY ORGANIZATION

One of the longest lasting debates in psychology is on the importance of the dispositional characteristics of the person versus the situation[1] in explaining behavior. This conceptual divide figures strongly in the differing accounts of how to explain personality. Today, most researchers in the trait tradition would agree that the dichotomy between person versus the situation is a

[1] The concept of "situation" in the personality literature often refers to the immediate situational context within which an individual is acting. Situation in the cultural and cross-cultural literature typically refers to the larger set of situations that individuals routinely encounter in their daily life. In empirical research, the focus has been typically on the broader set of situations that are prevalent in a specific environment, aligning more with the (cross-)cultural interpretation of situation.

straw man: they would agree that both stable dispositions and situational factors are important for understanding a person. Yet, the relative emphasis given to stable dispositions versus situational factors and the role that these factors play vary across theoretical accounts. Values historically have not strongly featured in this debate, although this is changing (Nilsson, 2014).

Dan McAdams (1995; McAdams & Pals, 2006) has provided one of the most interesting ways to integrate various types of personality concepts along the lines of what I included in Table 3.1. McAdams and colleagues proposed that personality is organized into a three-level hierarchy. At the most basic level are dispositional traits, which correspond to the Big Five traits. These are seen as dispositional and heritable. They are activated in social life and constitute the broadest, decontextualized, most stable, and recognizable aspect of psychological individuality.

At the second level are characteristic adaptations (McCrae & Costa, 1999), which include a wide range of motivational and social-cognitive aspects that are contextualized by time, place, and social roles. These encompass a smorgasbord of psychological processes, including motives, goals, plans, strivings, strategies, virtues, schemata, self-images, and mental representations. Importantly, values are included at this level.

McAdams and Pals (2006) assumed that characteristic adaptations and dispositional traits relate to each other interactively. How values and traits interact with each other has been debated. In the value tradition, Roccas and colleagues (Roccas, Sagiv, Schwartz, & Knafo, 2002) discussed three mechanisms that may explain linkages between values and personality traits. First, both may reflect some inborn temperaments; that is, individual differences in underlying biological predispositions are expressed in parallel traits and values. An individual thought to have a strong need for arousal may show higher trait levels of excitement seeking and may endorse Stimulation values, but not Security values. Second, there may be reciprocal influences between personality traits and values. Following classic balance theories in social psychology (Heider, 1946), individuals may reduce discrepancies between their cognitive system and behavior by either adjusting their values or changing their behavior. Hence, values may affect traits because individuals typically try to behave in line with their beliefs and values. By definition, values are guides for appropriate behavior and therefore may be used by individuals to modify and regulate their behavior. Alternatively, behaving in a consistent way (e.g., showing a particular behavioral trait) is likely to lead to a reassessment of what is important for a person (adjustment of values). If individuals perceive that they consistently behave in a particular way, they may modify their values to fit and justify their traits. A third position that contrasts these reciprocal arguments (McCrae & Costa, 1999) proposes that values are part of characteristic adaptations that are influenced by a continuing, stable core of the person (formed by the basic traits). Values as characteristic adaptations are the expressions of the basic dispositions (traits) in interaction with the social environment. Put more simply, values express personality traits while interacting with the social environment. Values are adaptive and culturally variable because they are the means for expressing invariant personality traits. The relative direction of the causal relationship between traits and values is one of the areas where current theories differ most from each other and will be the focus of the next chapter.

On top of this second layer of characteristic adaptations are integrative life narratives: stories created by individuals to construct meaning, purpose, and identity in their social and cultural world (McAdams & Pals, 2006). These narratives include reconstructions of the past and imaginations of the future that create a meaningful whole about that individual and his or her place in society. Life narratives build on the bottom two layers of basic dispositions and adaptive characteristics and integrate them into a social story that fits the social and cultural requirements of the environment within which an individual is living.

Two major forces shape the development and expression of personality in this tripartite hierarchy: genetic processes driven by Darwinian evolution and cultural processes. Genetic and evolutionary processes at the biological level are thought to be most important in shaping the dispositional traits, found on the hierarchy's lowest level. Culture in this theory has a weaker influence on traits, only affecting how they might be expressed in specific environments. For example, McAdams and Pals argue that Neuroticism in North America might be expressed in binging, excessive rumination, or ill-advised sexual adventures, whereas in rural Africa it might be expressed in somatic symptoms, social avoidance of perceived enemies, or engaging in magical thinking. Culture is thought to have the greatest influence at the highest level of the hierarchy, the life narratives, providing individuals with an extensive menu of how to create a meaningful life. Characteristic adaptations are situated at an intermediate level, being influenced both by biology and culture; they enable dispositions to be expressed within a context that is culturally shaped. Extending this rationale, Colin DeYoung (DeYoung, 2014) argued that traits should show higher heritability and lower cultural variability, whereas characteristic adaptations (which include values) show the opposite pattern (greater cultural variability and lower heritability). Hence, the relative influence of genes and culture on traits and values will give us a first indication of their position in this personality hierarchy.

As discussed in Chapter 2, the issue of inheritance is fundamental to evolution. There are two different forms of inheritance: genetic heritability, which relates to the relative influence of genetics; and social transmission and learning, which link to cultural influences. Determining the relative importance of genes and culture on traits and values is important because the answer will help us understand the relationship between the two.

HERITABILITY OF VALUES AND TRAITS

Heritability refers to the estimate of how much variation in a phenotype (e.g., Extraversion as a personality trait) is due to genetic variation among individuals within a specific population (Johnson, Penke, & Spinath, 2011). The most common way to estimate heritability is through twin studies, which compare similarity in phenotypes among monozygotic (identical) twins, who share 100% of their genes, to those in dizygotic twins and siblings, who share 50% of their genes. Twin studies separate the overall variation in phenotypes (expressed as personality traits or values) into genetic and environmental components, depending on how similar monozygotic twins are to dizygotic twins. The genetic component can be further divided into additive effects; that is, direct effects of each gene on a phenotype versus "interactive" gene effects. These include dominance effects (e.g., one allele is dominant over the other allele at the same gene locus, implying that only one allele version is expressed) and epistatic effects (multiple genes interact with each other in affecting a phenotype). Environmental components can be distinguished into shared (both twin have similar environmental influences) and nonshared environmental effects (everything else that is not shared by the twins). Heritability estimates can be statistically calculated, (but the interpretation can be complicated due to gene-environment interactions, which I discuss in later chapters).

Heritability estimates vary between 0 and 1, with 0 indicating that all phenotypic trait differences between individuals are due to environmental factors and 1 indicating that phenotypic differences between individuals are due to alternative genes that individuals hold within that specific sample. It is important to note that these heritability estimates are population dependent: in another population with a different constellation of environmental factors, the effect of genes may be different (Johnson et al., 2011).

Most research has focused on the relative heritability of personality traits. The relative estimates typically approach a heritability of 0.50 (Bouchard, 2004; Polderman et al., 2015) for personality in Western populations. The few studies that have been done in traditional societies and among hunter-gatherers in cultures that approximate our likely ancestral conditions suggest somewhat lower heritability estimates, but they are still of sizable magnitude (especially if heritability estimates are based on observer reports of personality traits and not on self-reports; Bailey et al., 2013).

In contrast, shared environmental effects are typically small in the personality trait studies that have been conducted to date. This suggests that the shared environment contributes relatively little to personality and what matters more are both genetic effects and the unique life experiences of each individual (which is captured in the nonshared environmental effect).[2]

Fewer studies have focused on the heritability of values. I review these studies in more detail than trait studies because many social scientists may not be aware of the extent to which values are influenced by genetic inheritance. In one of the earliest studies, heritability of work-related values in the United States varied between 0.18 (altruism values) and 0.56 (Achievement values); the average heritability across values was 0.38 (Keller, Bouchard, Arvey, Segal, & Dawis, 1992). In a Canadian twin sample (Schermer, Vernon, Maio, & Jang, 2011) that responded to a version of the Rokeach value survey (Rokeach, 1973), heritability varied between 0.36 (enjoyment, equivalent to Hedonism) and 0.63 (prosocial values; similar to Benevolence); average heritability across values was 0.46. In a study of Australian young adult twins answering the PVQ (Schermer, Feather, Zhu, & Martin, 2008), the average heritability was somewhat lower (averaged around 0.23) and varied between 0 (Achievement) and 0.38 (Tradition values). In a sample of German school-aged twins (Knafo & Spinath, 2011), estimated heritability varied between 0.34 for Openness to change versus Conservatism values and 0.49 for Self-Enhancement versus self-transcendence values. In a study of Israeli 7-year-old twins (Uzefovsky, Döring, & Knafo-Noam, 2016), heritability varied from 0 (Openness to change values) to 0.47 (Self-Enhancement values); the average genetic effect was 0.27. Shared environmental effects were close to zero, but nonshared environmental effects were substantive, suggesting that individual personal experiences are important. Finally, a study of Austrian twins (Renner et al., 2012) showed overall strong genetic effects. The additive genetic effects (indicating direct effects of genes on values) varied between 0.37 (for values resembling Self-Enhancement) and 0.65 for

[2] It is important to highlight again that most studies have been conducted as cross-sectional studies, with only one measurement point used to calculate the effects of genes versus environment. As I briefly mentioned earlier, heritability estimates can vary over the lifespan of an individual (or population). It is also plausible that longitudinal investigations provide more nuanced findings on the influence of the shared environment (e.g., the role of the family) on the development of stable personality traits (Schimmack & Lucas, 2010).

religion-related conservative values (average additive gene effects in models that included only additive and nonshared environmental effects were 0.44). Relatively strong nonadditive genetic effects (indicating that there are interactive effects between genes and dominant/recessive effects) were found for Self-Transcendence-related values. Shared environment again did not strongly contribute to values overall, similar to what was found in personality trait studies. The only value domain that showed shared environmental influence was in religion-related conservative values.

Overall, the heritability for values in Western samples was fairly similar to personality traits. Twin studies suggest that potentially up to half of the variability in both our traits and values is due to our genes. Although this finding might be surprising to some value researchers, most biologists are not surprised because most behaviors are heritable, even though the level of heritability is variable and may even change across the lifespan (Chabris, Lee, Cesarini, Benjamin, & Laibson, 2015; Turkheimer, 2000). What is important for my argument at this moment is that both traits and values are heritable to a similar degree.

GENETIC CORRELATIONS

Simple estimates of heritability, however, do not really address the issues of whether values and personality share some common biological predispositions. However, twin studies can be analyzed in a way that identifies whether values and personality correlations are due to shared genetic variance. In other words, we can construct genetic correlations between values and personality traits[3] that can give us insights about how genetic heritability is structured. This also addresses the interesting question of whether genetic correlations show the same dimensions as phenotypic correlations, which are typically analyzed in psychological studies. This analysis allows us to examine whether the genetic heritability is shared between conceptually related values and traits.

[3] This is typically done by applying complex multivariate statistics to twin data or large data sets of individuals whose genetic relationship is known (e.g., analyzing large numbers of family networks so-called pedigree analyses). It works similarly to heritability estimates discussed earlier. If two phenotypes are more highly correlated among monozygotic twins than among dizygotic twins (or among individuals who are genetically more closely related compared to more genetically distant relatives), then this information can be used to estimate the extent to which the two traits might be genetically related. As with heritability estimates in general, measurement artifacts (e.g., unreliability) and context effects are potential confounds that are difficult to control.

As discussed in the previous chapter, there is relative consensus about the dimensions and structure of values and personality when measured with self- and other-reports. Yet, it is unclear whether the same structures can be found when using genetic correlations. In one of the first such studies, McCrae and colleagues (McCrae, Jang, Livesley, Riemann, & Angleitner, 2001) reported an extensive analysis of the genetic structure of personality. Their study based on data from Canadian and German twins suggested that the genetic structure closely resembled the Big Five structure.

Focusing on values, the Australian study by Schermer et al. (2008) mentioned earlier broke down the value space into its genetic and environmental components. I conducted the same type of multidimensional scaling analysis that you are already familiar with from the previous chapter to examine how similar the two components are to each other. As a reference point, I included the raw correlations for values (averaging the positions for boys and girls, because Schermer and colleagues reported the correlations separately). I ran a separate analysis for each set of correlation matrices (genetic, nonshared environmental, and normal raw correlations of the self-reports) and then rotated the structures to maximal similarity to avoid capitalizing on random fluctuations. Figure 5.1 shows the joint space after rotation. As can be seen there, the prototypical value space was recovered quite well. The genetic and environmental components emerged in the same space as the normal self-reported values, with some minor deviations; for example, Openness to change values were somewhat more mixed, and Universalism values were placed in a wedge outside the Benevolence values. These changes are not that unusual, with Universalism and Benevolence often switching positions (see Schwartz et al., 2012 for a more extended discussion of this phenomenon). The overall value structure was well preserved. The important point is that both genetic and phenotypic (psychological) correlations show a similar structure in the personality and value domains.

This suggests that the same biological structure is involved, even though this does not necessarily imply that the same genes affect the same phenotypes to the same extent or at the same time of ontogenetic development (Johnson et al., 2011). To explain this seemingly contradictory statement, one gene might be related to each of two values, but then one of the values may also be influenced by a different set of genes while the other value is not. For example, a gene that makes people value relationships will be associated with Universalism, Benevolence, and Conformity values. Yet, another gene might make people concerned about upsetting others; in which case this gene is associated with only Conformity values, but not Universalism or Benevolence.

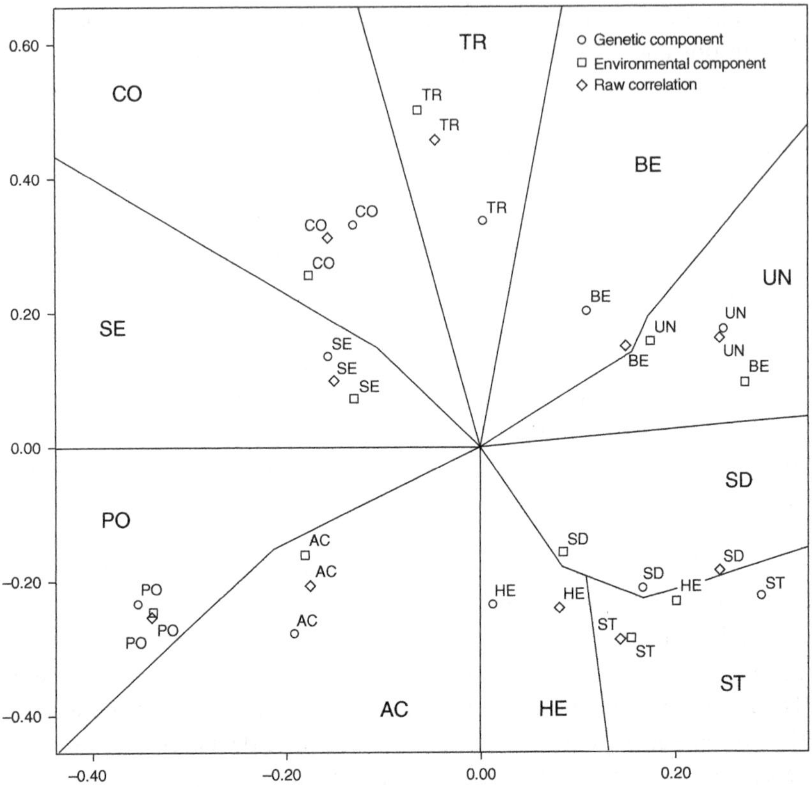

FIGURE 5.1 Two-dimensional organization of values using genetic, environmental, and phenotypic correlations. *Note:* The phenotypic correlations are analyzed separately for men and women; the genetic and nonshared environmental correlations are based on the total data set. PO = Power, AC = Achievement, HE = Hedonism, ST = Stimulation, SD = Self-Direction, UN=Universalism, BE = Benevolence, TR = Tradition, CO = Conformity, SE = Security.

Alternatively, the same gene might influence one personality trait or value early on during the development of an individual, whereas the same gene might influence another trait or value at a later point during development. For example, a gene for exploring the environment first affects Benevolence and Conformity because the immediate social context of an infant is the family and close family and friends, and it is this context that shapes the relationship that a child develops with these immediate contact persons. However, later the same gene might influence Openness to change values because the individual has left the immediate family environment, and the genetic effect is now directed toward exploration in a larger environment (which then influences other values that are of importance when interacting with a wider context).

The genetic correlation will be high in both cases, even though the underlying biological structure and processes are different. Nonadditive genetic influences (e.g., dominant alleles) or linkage disequilibrium (the tendency for two genes located in close spatial proximity to be inherited together) may also lead to spuriously high genetic correlations, even though there are no functional genes in common between phenotypes. Two values might be correlated at the phenotypic level because they happen to be driven by two different genes that are located close to each other on the same chromosome. I highlight these notes of caution because it is tempting to jump to easy conclusions based on these heritability studies. As I discuss in later chapters, how heritability translates into gene effects and biological pathways is a complex process that is not yet well understood.

Nonshared environmental influences may have somewhat different structures, especially when the effect of method bias (e.g., respondents answering a specific type of questionnaire) has been accounted for (McCrae et al., 2001). However, we saw earlier that for one sample of Australian twins answering a specific type of value survey, the nonshared environmental structure was identical to both the genetic and phenotypic (psychological) structure.

In summary, genetic correlations can indicate the high likelihood that either common genes are involved in the same trait or that there are reciprocal effects between genes and traits that may develop over time. Equipped with these insights, we can turn to the next challenge of whether personality and value genetic correlations are similar to phenotypic correlations.

GENETIC SIMILARITY OF VALUES AND PERSONALITY

The earlier mentioned study by Schermer et al. (2011) reported a study on Canadian twins in which both personality and values were measured. They used a version of the Rokeach Value Survey (Rokeach, 1973), which can be used to construct a simplified version of the two-dimensional structure described in Chapter 2.

Figure 5.2 shows the associations between this alternative value structure and personality with the labels used by Schermer et al. (2011). Let me quickly explain how this alternative structure maps onto the value labels that I have used so far. Enjoyment is equivalent to Hedonism; Achievement captures values from both Power and Achievement; Self-Direction is very close to Self-Direction as described in Chapter 3 (with some Stimulation values);

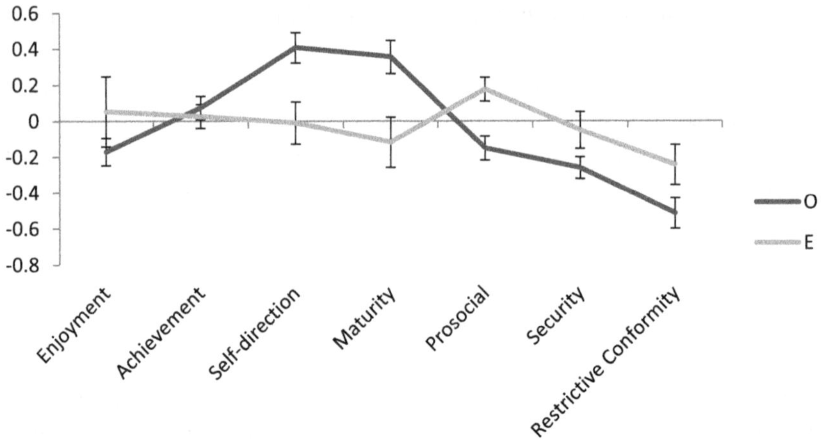

FIGURE 5.2 Genetic correlations between Openness and Extraversion traits and values. *Notes:* Enjoyment = Hedonism, Achievement = Power + Achievement, Self-Direction = Self-Direction (+ Stimulation), Maturity = Self-Direction + Universalism, Prosocial values = Benevolence, Security = Security, Restrictive Conformity = Conformity + Tradition.

Maturity captures a combination of Self-Direction and Universalism values; Prosocial values capture primarily Benevolence; and Security is similar to Security as defined and measured by Schwartz. Finally, Restrictive Conformity is a combination of Conformity and Tradition in the Schwartz model. The overlap is not perfect, but this earlier structure (Schwartz & Bilsky, 1990) already has the major differentiations of the later theory. To measure traits, a twin sample completed the Personality Research Form (PRF; Jackson, 2004), which can be recombined into the FFM.[4] I computed the average genetic correlations and the standard errors of the correlations based on the mean genetic correlations reported by Schermer et al. (2011). These correlations are plotted in Figures 5.2 to 5.4, and they correspond conceptually to the figures in Chapter 4 showing the raw correlations.

These correlations are based on the genetic component in the observed correlation that can be estimated by comparing the differences in correlations between mono- and dizygotic twins. In other words, if the genetic correlation is 0, then the influence of genes on the traits is independent of the

[4] The Personality Research Form scores can be recalculated to match the Big Five traits (Salgado, 2003; Stumpf, 1993). Salgado (2003) used expert ratings, whereas Stumpf (1993) reported a reanalysis of 18 cross-cultural data sets. I used these theoretical and empirical sources to cluster the 20 PRF dimensions into the Big Five. Two scales (Dominance, Dependence) that showed substantive cross-loadings (with differences in loadings of less than 0.05) were allocated to more than one of the Big Five traits.

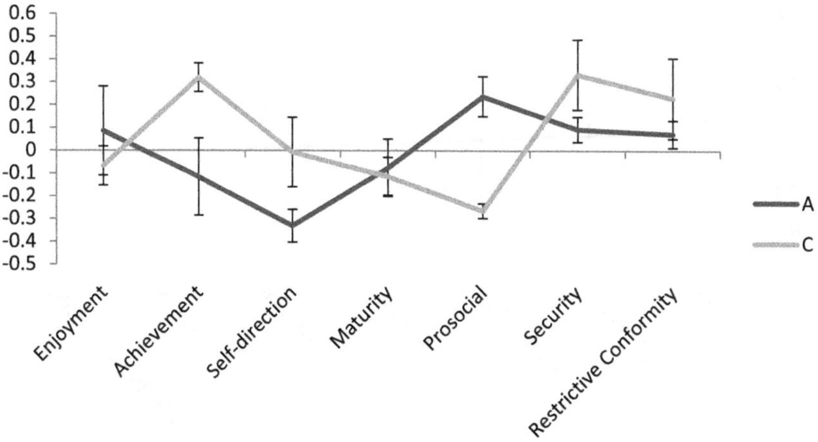

FIGURE 5.3 Genetic correlation between Agreeableness and Conscientiousness traits and values. *Notes:* Enjoyment = Hedonism, Achievement = Power + Achievement, Self-Direction = Self-Direction (+ Stimulation), Maturity = Self-Direction + Universalism, Prosocial values = Benevolence, Security = Security, Restrictive Conformity = Conformity + Tradition.

effect of genes on the values. If the correlation is 1, then the same genes are expressed in both traits and values.

The patterns reveal strong genetic bases for the correlations between values and personality. A number of patterns are noteworthy. First, Openness showed highly similar genetic and phenotypic correlations (based on the meta-analysis reported in the previous chapter). Self-Direction and Universalism values correlated positively, while Conservatism-related values correlated negatively with Openness traits.

Extraversion showed relatively weak genetic correlations across the various traits in the Schermer et al. (2011) study. Prosocially oriented values were positively related to Extraversion both in the meta-analysis and the genetic analysis, but the more complex associations with Openness to change and Self-Enhancement values were not found in the genetic analysis.

Agreeableness showed positive genetic correlations with prosocial values, which are similar to Benevolence values. This is similar to the phenotypic correlations in the value-personality meta-analysis reported in the previous chapter. The genetic correlations reached their most negative value for Self-Direction values, whereas for the meta-analytic raw correlations, the strongest negative correlations were observed for self-enhancing values. These findings are not entirely inconsistent because the earlier version of the Schwartz theory based on the Rokeach Value Survey lacked more

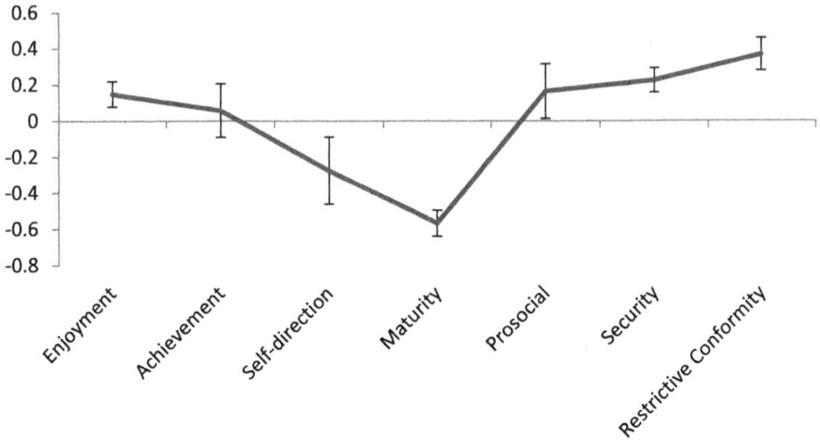

FIGURE 5.4 Genetic correlations between Neuroticism and values. *Notes:* Enjoyment = Hedonism, Achievement = Power + Achievement, Self-Direction = Self-Direction (+ Stimulation), Maturity = Self-Direction + Universalism, Prosocial values = Benevolence, Security = Security, Restrictive Conformity = Conformity + Tradition.

distinctive power and status-oriented values (Vauclair, Hanke, Fischer, & Fontaine, 2011). Therefore, the two traits that showed the clearest association with values in the meta-analysis (Openness and Agreeableness) also seem to share relatively strong shared genetic bases with values in this Canadian sample.

Conscientiousness showed relatively strong positive genetic correlations not only with Conservatism-related values but also with Achievement values. This mirrors the overall correlation pattern that we found in the meta-analysis, suggesting that these correlations might be driven by genetic effects.

Finally, the genetic correlations between Neuroticism and values revealed that a relatively strong genetic association underlied both constructs, which was much stronger than expected based on the phenotypic correlations in the meta-analysis. Conservatism-related values and Neuroticism showed strong positive correlations, which suggests that endorsing conservative values and emotional instability are encoded by the same genes. Maturity, which is similar to Universalism values, was negatively correlated with Neuroticism, which can be expected given the importance of ideals associated with Universalism that can induce anxiety and uncertainty (e.g., tolerating diverse perspectives and tolerance of other opinions). Enjoyment, a value type similar to Hedonism, correlated positively with

Neuroticism. Without trying to overemphasize this individual pattern, it may be consistent with some interpretations of Hedonism as a pathological mechanism to generate positive emotions to overcome emotional deficits.

Overall, the pattern of the genetic correlations is largely consistent with the overall correlation pattern reported in the previous chapter. Correlations between values and traits are likely to be driven at least in part by similar underlying genes. In contrast to what McAdams and Pals (2006) proposed, values and traits appear to be positioned at the lowest level of the personality hierarchy, both being biologically based universal personality characteristics. This is somewhat surprising given the great importance that values are typically accorded in the cultural psychology literature (Kitayama & Uskul, 2011). The work by McAdams suggested that genetic and cultural influences are inversely related; therefore, the next question is how strongly traits and values vary across cultures.

CULTURAL DIFFERENCES IN VALUES AND TRAITS

This section provides a first answer for one of the major questions of this book: how large are differences in personality traits and values between cultures? How different are the values and personality traits of Tina compared to those of Sandra, whom we both met in Chapter 1? How different are values and traits of an investment banker in Shanghai from those of a street vendor in Cairo, a hotel maid in La Paz, a business manager in Nairobi, a homeless person in London, or a teacher in Los Angeles? Newspapers, the popular press, and travel books are full of accounts that describe the alien other as having a different personality or valuing different things in life, but are they correct?

Newspapers when describing the behaviors of foreign leaders or populations often implicitly touch on their personalities and values when explaining the events to the audience at home. A whole industry exists around training managers, politicians, students, or tourists to recognize and help them understand the stereotypical behaviors, traits and values that characterize one nation compared to another. Yet, the question of the size of any such cultural differences in values and personality traits has received relatively little attention until very recently.

In contrast, self-descriptions by groups of people from different cultures often overlap in quite interesting ways. Here is one random example: during a visit to the Dubai Museum in the United Arab Emirates, I found a display that described the traditional Bedouin dwellers of the Arabian peninsula using a mixture of value and personality traits: Bedouins love good deeds and hate evil; they are very sociable and hospitable by nature, generous, respectful, and patient; they have strong personalities; they protect their neighbors; and they are happy. Thousands of kilometers away, a list of Norwegian Viking rules in a spectacular outdoor museum set among waterfalls and eerie mountains included similar personality and value descriptions, emphasizing honesty, loyalty, hospitality, and courage. Two peoples of very different ecological and historical traditions – one group living in one of the hottest and driest climates, the other living in a Nordic cold and wet sea climate – share very similar self-descriptions. Which of these accounts is more close to the empirical facts? As discussed earlier, this question of cultural variability has some interesting theoretical implications because it can provide insights into how human personality is organized.

Effect Size and Explained Variance

To explain how big cultural differences are, I need to briefly touch on two related concepts: effect sizes and explained variance. Effect sizes are simply what the name suggests, an estimate of how big a difference exists between groups. There are a number of effect size types that examine, for example, the size of group differences, or the relative strength of an association between two or more variables (correlations), or risk estimates (whether one treatment is more or less effective in curing a disease). For our purposes, effect sizes that estimate differences between two or more groups are important. The simplest case is based on two-sample comparisons such as Cohen's d, but multisample cases that compare more than two cultures or societies include measures such as eta squared, the intraclass correlation, or R squared (Cohen, 1988, 1992; Ferguson, 2009). Although used in different statistical methods and traditions, they use the same principle of variance and typically can be transformed from one form into another in a relatively straightforward manner.

The basic idea of these effect sizes is that any variability in observed scores can be broken down into variability that is due to some systematic effect of group membership (in our case, culture, ethnicity, or nation) and variance that is not explained by this group effect (called error variance). The variance of interest and error variance add up to the total variance; that is, the total variability in the observations. Effect sizes are then often expressed as the ratio between the variability that is due to the grouping of interest (culture, ethnicity, nation) and the total variance. Effect size estimates vary between 0 and 1, with 0 indicating that the particular grouping has no meaningful effect and 1 indicating that the effect of group membership predicts all the variance of the observations. If multiplied by 100, these estimates can be expressed in percentages of explained variance. An effect may explain 0% of the variability or 100% of the variability in scores. In most real-world cases, the effect size is somewhere in between. In an extreme case, all the variability in a particular trait is due to one's cultural background. For example, if you were born in Denmark, then you are very unlikely to speak Chinese fluently. In contrast, an individual born in Taiwan will naturally speak a dialect of Chinese. Hence, the variability in speaking Chinese between two samples is nearly 100% due to one's cultural background (not counting the possibilities of migrant families raising their children as bilingual). For an example of close to 0% cultural variance, the ratio of males to females at birth is likely to be unrelated to culture[5]: the chances of giving birth to a boy or girl are pretty much random and independent of where you are living. This leads to a variance estimate that is close to 0.

There are a number of cut-off criteria to decide whether an effect is small, medium, or large. Unfortunately, these values are often arbitrary, representing intuition, experience with the data, conventions, or empirical distributions (Cohen, 1988). Ferguson (2009) suggests a recommended minimum practical effect size (RMPE) as a lower criterion for a "practically" significant effect in the social sciences. For the variance-based effect sizes that are of interest for us here, the recommended minimum for interpretation is 0.04 (or 4%). Moderate effect sizes are 0.25 (or 25% of explained variance), and strong effect sizes are larger than 0.64 (explaining 64% of the variance). These criteria are only guidelines, and different disciplines and research areas may adopt different cut-off criteria. For example, if the outcome variable of interest is dichotomous and has practical relevance (e.g., life or death), even small effect sizes of 0.01 can be extremely important because it means that an intervention can save one life in 100 cases. However, if variables are measured using rating scales with substantial measurement error and poor standardization, these criteria may need to be adjusted upward.

It is also important to consider what these estimates imply for our ability to decide whether a person belongs to one cultural group or another. Figure 5.5 shows two different normal distributions of traits. On the left

[5] This statement is not really true anymore. Advances in medicine enable us to determine the sex of a fetus, and the ability to abort relatively safely has radically changed the sex ratio in some populations. In some cultures, boys have higher status and are more desirable for families, which unfortunately has led to a high rate of abortion of females. Therefore, cultural norms together with modern medicine have shifted the natural birth ratio.

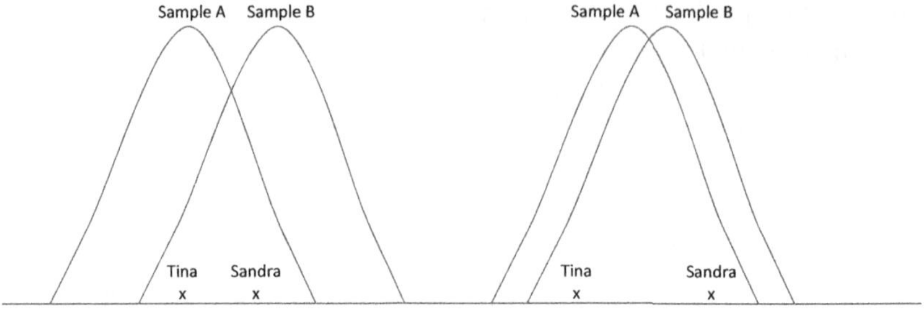

FIGURE 5.5 How easy is it to identify individuals based on underlying distributions?
Note: The figure shows two normal distributions of a single underlying trait. On the left the two populations show a moderate degree of difference, while on the right the two distributions show only small mean differences. Yet, in both cases it is empirically difficult to determine whether any individual belongs to one or the other distribution.

side, you have a relatively large difference (this graphs shows a difference that would show up as moderate to large in an empirical investigation – the two distributions are clearly distinct in their means). On the right side, you see two distributions that are closely overlapping, showing a small difference. Imagine we wanted to locate Tina and Sandra within these two populations. In both cases, we would be unable to determine to which group they belonged, even though the differences in their scores are large (and in the case of the right set of distributions, they are located at nearly opposing ends of the underlying distribution). Even large variance estimates on a single trait may not allow us to decide easily which population an individual member is likely to belong to (in statistical terms).

Effect Sizes for Trait Differences across Nations

Relatively few studies have reported the effect sizes for cultural differences between countries. The largest personality study (Bartram, 2013), whose sample was made up of employees applying for jobs (instead of students), showed relatively modest cultural differences. The largest variance explained by culture was estimated for Emotional Stability (0.14), followed by Extraversion and Agreeableness (both 0.11) and Conscientiousness (0.08); the least amount of variance explained by culture was in Openness (0.06).

An earlier meta-analysis of the Eysenck Personality Questionnaire (EPQ), which provides scales for Extraversion and Neuroticism that are comparable to Big Five instruments showed about 17% of explained variance by culture in Extraversion and about 16% in Neuroticism (Hemert, de Vijver, Poortinga, & Georgas, 2002).

Schmitt and colleagues (Schmitt, Allik, McCrae, & Benet-Martinez, 2007) conducted a large multinational study involving students samples in 56 nations. They used the Big Five Inventory (BFI; John et al., 2008), which I mentioned in Chapter 3. Schmitt and colleagues reported eta-square values, which according to Cohen (1988) can be classified as small if 0.01, medium if 0.06, and large if above 0.14. The smallest effect size was found for Extraversion (0.03), followed by Neuroticism (0.05) and Openness (0.07). The largest effects were found for both Agreeableness and Conscientiousness (both =.09). Therefore, these estimates were overall of medium size, indicating that there is moderate variability in personality trait levels across national samples.

McCrae and colleagues (2005) conducted a study with the NEO-FFI, where they asked students to observe and rate the personality of two individuals within their culture whom they know well. On average, about 4% of the variance was explained by the nationality of the participants, with the largest amount of variance being found for Openness (5.1%) and the least amount for both Neuroticism and Conscientiousness (3.4%) (McCrae et al., 2005). Therefore, their effect sizes (adjusted for age and gender differences) were of small magnitude.

The most recent wave of the World Value Survey included data from a very short 10-item measure of personality traits (Rammstedt & John, 2007) administered to representative samples across 18 nations and regions (N = 23,053). I ran a one-way ANOVA to estimate eta-squared values. The average explained variance was 18.2%. The individual estimates varied between a low of 5.7% for an Openness item ("have few artistic interests") and 9.9% for an Emotional stability/Neuroticism item ("gets nervous easily") to a high of 29.5% for an Agreeableness item ("is generally trusting") and 38.2% for a Conscientiousness item ("does a thorough job"). Combining the variance estimates for the two measured items per trait, the largest variance across nations was for Conscientiousness (29.8%), and the lowest variance was observed for emotional stability (10.0%). Agreeableness (20.7%) and Extraversion (18.1%) also showed considerable variance across these 18 nations. This again suggested moderate to large differences.

Overall, this evidence does not fit well with the suggested relative independence of personality traits from culture, if we assume that traits form the biologically driven lowest level of the personality hierarchy. Let us examine next how values fare in this respect: do they show more or less variability across societies?

Effect Sizes for Value Differences across Nations

A few years ago, I conducted a large reanalysis of various value data sets with Shalom Schwartz (Fischer & Schwartz, 2010). In our first study, we examined the variability in responses to the Schwartz Value Survey (SVS) by teachers and students in 67 countries (N = 41,968). Among teachers the average explained variance by nation was 12.5%, whereas among students it was 10.9%. These estimates varied between a high of 19% for Conformity values and a low of 6.0% or 8.0% percent for Self-Direction values among student and teachers, respectively.

In the second study, we analyzed data from a short version of the PVQ that was administered to representative samples in 19 countries in Europe (N = 42,359). Overall, the estimated cultural variance was 4%, varying from a low of 2% (Stimulation) and 3% (Universalism) to a high of 6% (achievement) and 10% (Hedonism). As could be expected, variability across Europe is lower because cultural, social, and economic conditions are more similar in many European countries than across a larger sample of countries from around the world (I return to this issue in later chapters).

In an as yet unpublished study, I examined the variability in single items measuring the 10 value types in a survey resembling the PVQ that was administered across representative samples from 56 countries (N = 71,916) as part of the World Values Survey (see Welzel, 2010, for more information on these data). Overall the explained variance by nation was similar to the previous estimates using the SVS (12.5%). The individual estimates ranged from a low of 7.2% for Universalism values to a high of 20.3% for Power values.

These estimates were based on various versions of instruments developed by Schwartz. Do values measured using different approaches show smaller or larger differences? In our final study together (Fischer & Schwartz, 2010), we also included items measuring values following the Inglehart and Welzel tradition in representative samples from 62 nations and world regions (N = 84,887). These value items capture value-driven behaviors or attitudes (as discussed in Chapter 3). The overall variance due to nation was slightly larger than in the other studies, averaging 22% across all items. It ranged from a low of 16% for how happy people felt to a high of 30% (for how important God is in the life of the respondent).

What is interesting to note is that the variance estimates for values are of similar magnitude as for personality traits, and sometimes even a bit smaller. This is somewhat surprising given the supposedly stronger effect of culture on values compared to personality.

Summary of Cultural Variance Estimates for Traits and Values

Both personality traits and values show moderate levels of variability across national samples. The specific estimates vary by type of samples and type of measurement. Among personality traits, Agreeableness and Conscientiousness showed relatively large variability in trait levels across nations (except in the observer ratings). Extraversion and Neuroticism sometimes showed large differences (in particular in employee samples), but in student samples there was relatively little variability for these two traits. We could speculate whether samples made up of psychology and social sciences students are more homogeneous across the world than employee samples.

For values, there was a relatively consistent pattern of greater variability for Conservatism and Self-Enhancement values across nations and lower variability for Self-Direction, Stimulation, and Self-Transcendence values. The greatest variability was observed for religion-related values ("God is important"). This dovetails with some other evidence suggesting that religion and political variables show greater differences across nations than do values and personality traits (Saucier, Kenner, et al., 2014). However, the overall estimates for values were within the same range. In the one database where both values and personality traits were administered with short scales (the World Value Survey), personality traits on average showed somewhat higher variability than values. This suggests that, as far as cultural differences are concerned, traits and values are located at the same level of the personality hierarchy.

GENES VERSUS CULTURE: REVISITING THE PERSONALITY HIERARCHY

We have covered a lot of material in this chapter. The hierarchical structure of personality proposed by McAdams and Pals (2006) put traits at the lowest level of the hierarchy; they are driven by evolutionary pressures of living in social groups. Traits are considered the most basic form of personality, are strongly influenced by biology, and have only limited variability across cultures. Values, in contrast, form part of characteristic adaptations, which are the next layer of the personality hierarchy. Values and traits are thought to be interlinked, but the influence of genes on values is considered to be weaker and the influence of cultural processes is deemed to be stronger. On top of these two layers, McAdams and Pals place life stories and culturally influenced narratives that people create as they make sense of their day-to-day experiences in their specific social, economic, cultural, and historical context. The life stories of an extrovert living in the Middle Ages in Venice will be very different from the life stories of a modern Italian extrovert.

Recent evidence suggests that this relative ordering of values and traits may not be empirically tenable. Most importantly, studies have found that the heritability of both traits and values is substantial. The influence of genes turns out to be much greater than has been assumed in some domains of the social sciences. Biologists and personality researchers might be less surprised by these estimates. Despite widespread claims in the personality and social psychological literature that values are more strongly determined by culture than by genes, the data show that genes have about an equally important influence on values and on traits. Our genes do matter a lot for who we are.

A second and clear pattern to emerge from recent data is that values and traits are influenced to significantly smaller, but similar degrees by national differences. Returning to Tina and Sandra, the general assumption among social scientists is that these two individuals are likely to have very different traits and values, because they were raised in two very different cultural contexts. You may have thought the same when reading the brief descriptions of them at the beginning of this book. Yet, the evidence suggests that underlying the very different behaviors that both engage in, their traits and values at the psychological level are actually much more similar than the exotic surface differences may suggest. Both young women value their close family. Tina deeply appreciates having her family support her in her duties as a spirit medium, whereas Sandra is proud of her brother and fondly recalls the most recent family dinner, which was really important for her. Both Tina and Sandra are also very meticulous and conscientious in the tasks they perform in their respective cultural environments. These two strong women are more similar than the very different actions that they take would make us believe.

To put it differently, whether one is a Thai or a Brit, or a Thai-Chinese or a true Londoner matters much, much, much less in determining one's personality than one's genes. If you think of the levels of influence as slices of a cake, potentially up to half of the cake is taken by genes. Only one meager slice of the cake can be claimed by culture. This is quite an imbalance that may surprise you. Although our environment, which includes culture, does matter (which I discuss in Chapters 8 and the following chapters), in the direct head-to-head competition between genes and culture, genes take the lion's share.

A third pattern that is emerging is that phenotypic correlations reviewed in the previous chapter and genetic correlations observed in twin studies are surprisingly consistent.

All these lines of evidence converge on the same two conclusions: traits and values are much more similar than many people would assume, and both values and traits are linked via common genes. My colleagues in social psychology, and especially in cultural programs such as anthropology and sociology, often stress the importance of social and cultural processes. They see socialization as fundamental for understanding who we are as a person. This view needs updating in light of this evidence coming from large-scale and representative samples from all inhabited continents. Genetics matter and socialization much less so for contributing to the formation of personality and values.

This conclusion leads to an updating of the personality hierarchy: traits and values probably form the lowest layer of the personality hierarchy together. Characteristic adaptations – that is the specific plans, goals, strategies, self-images, and developmental tasks that are shaped by our environment – are informed by both traits and values, which are more strongly influenced by our genetic heritage than by our cultural socialization.

SUMMARY

In this chapter, I took a first stab at answering the question of how important our genes and our cultural environment are in shaping our personality traits and values. The personality literature has long assumed that traits are fundamentally biological and therefore show little variability across cultures but relatively high genetic heritability. Both value and personality researchers also assumed that values are much more strongly influenced by culture and only marginally influenced by genes. Yet, the empirical evidence suggests that both values and traits are strongly determined by genes (as estimated in twin studies), with culture playing a weaker role. Studies examining the correlation between values and personality traits also suggest that the associations that we had seen in the previous chapter are strongly determined by our genetic make-up.

The interesting question that emerges now is what mechanisms – both genetic and in the brain – are common to both values and traits. The two traits that are most similar to values are Agreeableness and Openness traits; they converge quite strongly with Self-Transcendence (versus Self-Enhancing) and Openness to change (versus Conservatism) values, respectively. Conscientiousness and Extraversion (and also Neuroticism), in contrast, seem to have a more mixed motivational profile. What are possible mechanisms that underlie these associations? This is the challenge for the next chapter, in which I review various theories of traits that can provide some insights.

6

Searching for the Underlying Mechanisms in the Brain and the Situation

In Chapter 2 we saw that values and traits are quite systematically correlated. In the previous chapter, we found that values and traits are both strongly inherited and show relatively weaker cultural variability. In other words, their mean levels do not vary much across societies. In this chapter, I review different theories explaining the underlying architecture of the Big Five. Because there is no comparable work in the value domain, I focus on these personality theories and then broaden the discussion to include values toward the end of the chapter.

My goal is to get a better understanding of the genetic and neuroscience processes involved in both values and traits. As we will see, there are many theories, and the relevant evidence is sometimes mixed and contradictory. Please bear with me as we wade our way through a sea of theoretical details in search for clarity. The world is complex, and so is personality. Together, let us tackle this task of making sense of 40 years of diverse strands of research into genetic and neurocognitive mechanisms that explain why our personality is structured the way it is.

FRAMING THE DEBATE: THE (NEUROBIOLOGICAL) PERSON VERSUS THE SITUATION

We have made much less progress in understanding WHY we have this particular trait structure as captured in the Big Five than in describing this structure. In all fairness, the five-factor model has only been around for about a quarter-century, and major breakthroughs in genetics and neuroscience that would enable an investigation of how such traits might have emerged have only occurred even more recently. Yet, strong foundations for personality research were laid much earlier (Cloninger, 1986; Cloninger, Svrakic, & Przybeck, 1993; Gray, 1970, 1975). As we had seen in the previous chapter, both genes and culture as one form of a situational context play a role. The debate over genetic versus environmental influences has been ongoing since Mischel's famous critique of personality research (1973, 1977). Put simply, there are two major explanations for the emergence of personality trait structures: theories that give priority to genetic and neurobiological systems versus those that give priority to situational characteristics as perceived via social-cognitive mechanisms. There is also an intermediate position that links both explanations via so-called neural network structures. The ability to account for all five (or six) major traits in the personality literature is challenging; in particular, Agreeableness and Openness traits are less well predicted. For this reason, it is worth exploring neurocognitive models that have tried to explain related processes, most notably prosocial behavior and empathy. I review these major sets of theories and supporting evidence first and then present an integration of value-trait relationships in the final section of this chapter.

NEUROBIOLOGICAL THEORIES OF PERSONALITY

Reinforcement Sensitivity Theory

The most important neurobiological model is the reinforcement sensitivity theory first outlined by Jeffrey Gray (1970). Its goal was to explain the underlying biological and psychological dynamics of Extraversion and Neuroticism, the two major traits of personality then under discussion (Eysenck & Eysenck, 1965, 1975). The theory has undergone substantive revisions over the years, and I focus here on the most recent variations and their application to the Big Five traits (Corr, 2004, 2008; Corr, DeYoung, & McNaughton, 2013). Values are not of primary concern for this theory.

Reinforcement sensitivity theory begins from the assumption that behavior is motivated by obtaining rewards and avoiding punishment. It follows cybernetic systems approaches that focus on the survival and reproduction of biological systems (Carver & Scheier, 1982, 2002). Cybernetics assumes that systems operate based on goals and feedback: a system works toward a specific end-goal and changes its operations based on feedback received that indicates how the system is progressing toward achieving that goal. Figure 6.1 shows a basic cybernetic feedback cycle.

Reinforcement sensitivity theory combines these cybernetic feedback loops with ideas from behaviorism that emphasize rewards and

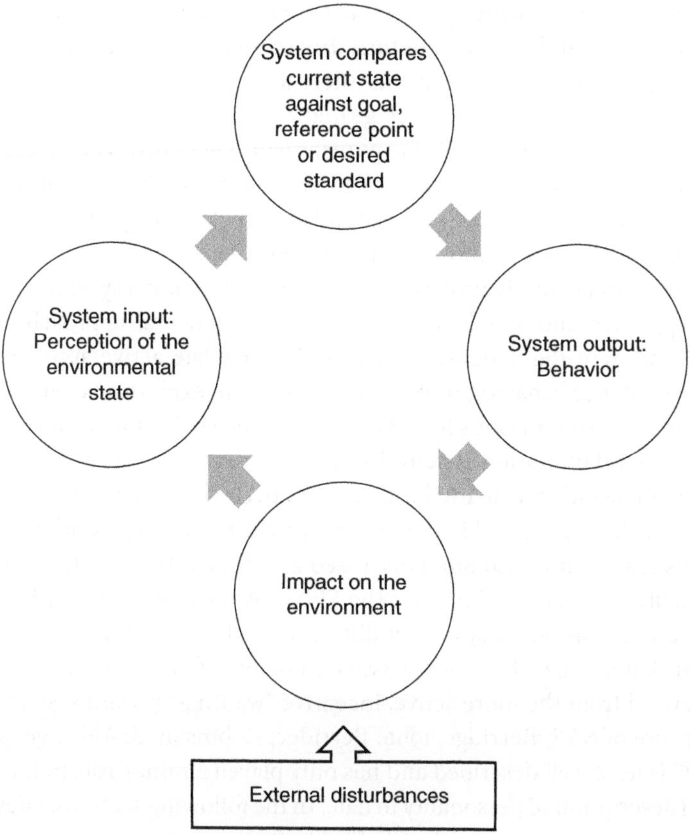

FIGURE 6.1 A basic cybernetic feedback cycle.

punishments. Rewards are end-goals toward which a person works. This striving toward rewards then leads to a higher frequency of behaviors that help the system obtain the rewards. Such rewards can be obvious things of importance for the survival of any living being, such as food, drink, or sexual mates, but they can also be less obvious, such as the termination or avoidance of imagined or real unpleasant states (e.g., pain, denial of food, rejection by potential mates). In contrast, punishment is associated with aversive stimuli (e.g., unpleasant body states, predators, poisons), which any biological system is motivated to avoid. This avoidance in turn leads to a decrease in behaviors associated with these stimuli. Punishments may also include imagined or expected states (e.g., fear of something) or the denial or omission of an expected positive reward (leading to frustration). Rewards and punishment are the primary forces that are used to evaluate inputs coming into the system and that then in turn drive system output (behaviors).

Gray (1970, 1975) proposed that there are two major classes of behavioral output: approach behaviors geared to attaining rewards and avoidance behaviors geared to avoiding punishment. These two types of behavior (approach vs. avoidance) are controlled by a number of theoretically derived motivational systems, which are grounded in biological systems (the brain systems that give rise to those motivational systems).

The theory currently proposes four major motivational systems that regulate approach and avoidance behavior. The behavioral approach system (BAS) and fight-flight-freeze system (FFFS) regulate active approach and active avoidance behavior, respectively: is it safe to explore an environment, or should one run for one's life? Passive avoidance behavior is controlled by the behavioral inhibition system (BIS). This system is concerned more with the general avoidance of anything new or novel when there are any doubts about its safety. It is an "if in doubt – better safe than sorry" kind of system. For this reason, it is also often discussed as a system that resolves conflicts (especially between the BAS and the FFFS). A pleasure system (PS) is the most recently proposed system: it differentiates between a hedonic "liking" system characterized by a more passive enjoyment following the attainment of a reward from the more active, incentive "wanting" reward system (seen as the core of BAS; Berridge, 1996; Berridge, Robinson, & Aldridge, 2009). The PS is least well described and has only played a minor role in the theoretical description of personality to date. In the following sections, I describe these systems in relation to the Big Five traits.

Avoidance Motivation and Neuroticism

The FFFS is primarily an active avoidance system that comes into play in the face of threat. It is activated in response to aggressive attacks and may involve freezing and panicking or cautious exploration of, rapid withdrawal from, or active avoidance of unpleasant situations and stimuli. What reaction is chosen depends on how close the aversive stimuli is perceived to be (e.g., can I still get away or is my only option to attack?). It also depends on the overall sensitivity of the system to risk and the resultant level of anxiety. Some people might be more likely than others to approach an unknown object with an unknown risk. This overall sensitivity in turn is regulated by the BIS.

The BIS becomes particularly important when there is a goal conflict. Most common are approach-avoidance conflicts (e.g., wanting to talk to a potential mate but fearing rejection), but other conflicts are also possible. For example, an approach-approach conflict arises when there is a need to decide between two desirable options (e.g., two attractive job offers). The complex interactions of the BIS and FFFS are thought to be associated with Neuroticism. Individuals who rate high on Neuroticism are unable to efficiently resolve such conflicts and therefore experience uncertainty, anxiety, and mood swings. The operation of these two systems is also used to explain the two different components of Neuroticism (volatility vs. withdrawal). Neuroticism can be associated with aggressive and violent impulses if a person feels threatened and sees no other form of reacting to the threat.

The major neurotransmitter that has been associated with behavioral avoidance motivation in general (and with Neuroticism in particular) is serotonin (Collier et al., 1996; Lesch et al., 1996). Medication to treat major depression often targets the active role of serotonin in regulating Neuroticism-related states such as anxiety and depression (Hieronymus, Emilsson, Nilsson, & Eriksson, 2015). There is now little doubt that serotonin plays a major role in regulating mood and anxiety, which are linked to behavioral avoidance motivations.

Not surprisingly, individual differences in Neuroticism (and more broadly in behavioral avoidance motivation) have been found to be associated with brain differences in size and connectivity in areas that regulate both positive and negative emotions (e.g., the cingulate and the amygdala), memory (e.g., the hypothalamus), and complex decision making (e.g., in the prefrontal cortex). This brain network system is functionally associated with activation in the hypothalamic-pituitary-adrenal (HPA) axis, which regulates a person's stress response, including the body's primary and most well-known stress hormone, cortisol.

Approach Motivation and Extraversion

The primary function of the behavioral approach system (BAS) is to motivate a system to move toward achieving its desired goal state, principally by approaching perceived rewards. If we are conditioned (we have learned that something is desirable or undesirable) to perceive something as positive, we are motivated to experience the associate positive feeling again. We want to be with our loved ones, we are keen to eat the food that we really like, we want to do the things that make us feel good (which can be bad for us in the long run – think of the negative effects of drugs). Even unconditioned things can be rewarding. If you felt extreme hunger or thirst, you have experienced an unconditioned response that strongly activates the biological system. The resultant approach behavior is associated with desire, excitement, and hope. The pleasure system (PS) is less clearly described, but has been implied through pharmacological studies of neurotransmitters such as dopamine and various opioids (Berridge et al., 2009). Opioids are chemicals (peptides) produced by the body that are centrally involved in regulating mood (making us euphoric, elated, depressed, or sedated), feelings of pain, and appetite. The PS is activated by the attainment of a reward and produces states of enjoyment, cheerfulness, and satisfaction. Achieving a goal often requires complex hierarchies of subgoals, which require (1) identification of appropriate goals at each level of a hierarchy (e.g., to get good grades in this course, I will need to study for the test next week); (2) plans to achieve a goal (e.g., I should read all the articles on the reading list); and (3) execution of the plans (e.g., setting time aside in the next hour to actually do the readings). These goal hierarchies are not thought to be formally part of the BAS (or BIS) by Corr and colleagues, but can be regulated and influenced by motivational states that are controlled by either the BAS or BIS (Corr et al., 2013).

Extraversion is the personality trait most directly associated with the BAS, and reward sensitivity is often argued to be the major driving force underlying Extraversion. Although Extraversion has strong social connotations (e.g., assertiveness), humans are ultrasocial. Therefore many rewards are social in nature (e.g., status, attracting mates), and we humans use social strategies and behaviors to obtain potential rewards (e.g., to get a promotion, I need to motivate colleagues to work with me on projects that bring positive results that count toward meeting my performance criteria). In fact, people high in Extraversion may perceive social situations as rewarding in themselves (Wilkowski & Ferguson, 2014). In light with the reward sensitivity that is thought to underlie Extraversion, neuroscience research has shown some associations with major brain reward systems and areas involved in complex decision making; for example, the ventromedial prefrontal cortex (VMPFC), the striatum, and the anterior cingulate cortex (ACC) (Adelstein et al., 2011; Allen & DeYoung, 2015; DeYoung, 2010).

One of the major neurotransmitters that is thought to engage the reward system is dopamine (DeYoung, 2013; Wacker & Smillie, 2015; Wacker, Chavanon, & Stemmler, 2006) because it is centrally involved in regulating reward sensitivity. Therefore, the stereotypical account pits serotonin as regulating the behavioral avoidance systems against dopamine regulating the behavioral approach system, leading to a tug-of-war relationship between these two neurotransmitters. Depending on the balance of these neurotransmitters in the neural system of a person, the general tendency is shifted either toward behavioral approach or avoidance motivations, associated with variations in Extraversion and Neuroticism, respectively.

Beyond Extraversion and Neuroticism

Reinforcement sensitivity theory initially focused on explaining the underlying biology of Extraversion and Neuroticism. As a consequence, most of the theoretical discussion and empirical evidence relate to these two major traits. The other traits have received considerably less attention, and the evidence is relatively mixed. In this section I briefly present the major research for the other three traits in the Big Five. You may wonder about the HEXACO model. As you may recall, in Figure 4.7, Agreeableness and Honesty-Humility emerged in a single cluster. This finding is also in line with other recent studies that suggest that Honesty-Humility might be part of a broader Agreeableness trait (De Raad et al., 2010, 2014).

Openness, Curiousity, and Behavioral Approach Motivation

As discussed in Chapter 2, Openness as a trait is associated with a predisposition toward intellectual curiosity, interest in innovation, unconventionality, and fascination with more complex and philosophical thoughts. This interest in novel and unconventional things is conceptually related to the BAS and brain reward systems (see Allen & DeYoung, 2015; Corr et al., 2013). Receiving new information that one is curious about may be as rewarding as receiving money or social rewards. Extraversion and Openness to change differ in that the former is more related to tangible and socially focused rewards, whereas Openness is more related to information and cognitive engagement and the rewards that stem from engaging with these activities. There is also indirect evidence that dopamine is involved in behaviors that underpin or facilitate processes that are important for Openness (e.g., curiosity, exploration, work memory and attention, and latent inhibition; Allen & DeYoung, 2015).

Curiosity is a particularly interesting part of Openness. It is one of the most important evolutionary motivations that encourage organisms to seek out novel stimuli and information and that allows learning to take place (Kidd & Hayden, 2015). Being able to move around (compared to plants), organisms need to explore their environment and integrate information about safe food resources, potential allies, and threats (predators, poisons, etc.). Obtaining information is also rewarding in itself and activates reward systems in the brain. For this reason, curiosity is likely to be associated with approach motivation regulated by the BAS. As mentioned, Berridge and colleagues (Berridge, 1996; Berridge et al., 2009) have suggested that there are two distinct motivational approach processes: "wanting," associated with incentive salience, and "liking," which is associated with hedonic impact and enjoyment. These two processes are associated with different neural mechanisms. Wanting is related to dopaminergic brain systems that are widely distributed. In contrast, liking involves fewer areas and these are regulated by opioid systems. They are the systems that Corr et al. (2013) proposed as differentiating the BAS from the PS (pleasure system). Most likely both systems are involved in Openness (but see Litman, 2005).

Conscientiousness: The Avoidant Nerdy Trait?

Conscientiousness is often considered the most complex and speculative trait in terms of motivation (Corr et al., 2013). As I discussed in Chapter 2, Conscientiousness is associated with being organized, reliable, self-disciplined, orderly, and hard-working. Its two major aspects are industriousness and orderliness. Industriousness captures the the pursuit of nonimmediate goals and the control of possibly disruptive tendencies in pursuit of those more distant goals, whereas orderliness refers to the tendency to adopt and follow rules. Hence, Conscientiousness is both approach oriented in the pursuit of distant goals (industriousness) and avoidance oriented in avoiding punishment resulting from violating rules and impulsive tendencies that may interfere with long-term goal attainment (both industriousness and orderliness). As discussed earlier, Neuroticism is the trait most clearly associated with the BIS. Conscientiousness should show therefore some positive relationship with Neuroticism. However, as we saw in Chapter 3, the overall relationship is negative at the trait level. More conscientious individuals are typically more emotionally stable. A study separating the variability within individuals over time and variation between individuals showed that Neuroticism and Conscientiousness are positively related within a single individual, but are negatively related when comparing overall trait levels between individuals (Beckmann, Wood, & Minbashian, 2010). This means that while working on completing a goal-directed task, individuals may experience anxiety (a key marker of BIS) because possible punishments may stem from not completing the task. At the same time, those individuals who are working hard and complete tasks on time are less likely to experience punishment, which makes them less likely to have increased levels of anxiety and Neuroticism.

A Bit of Speculation: Agreeableness

The underlying motivation of Agreeableness is equally opaque. Agreeableness is oriented toward compassion (understanding how other people are feeling, attachment and concern for others) and politeness (suppression of aggressive or norm-violating impulses, especially those that might negatively others). Prosocial behaviors are intrinsically rewarding (Declerck, Boone, & Emonds, 2013), suggesting that Agreeableness might be partially motivated by approach motivations. At the same time, constraining aggressive and norm-violating impulses is clearly associated with avoidance motivation. Hence, both approach and avoidance motivational systems are likely to underlie Agreeableness.

A Brief Summary of Reinforcement Sensitivity Theory

This theory is arguably one of the most widely cited theories for explaining how the Big Five traits function at a neurobiological level. Its starting premise is that people operate like cybernetic systems, reacting to environmental stimuli and being motivated in their actions to seek rewards and avoid punishments. These motivations then give raise to two broad classes of behaviors: either approaching or avoiding situations. These two classes are guided by a small number of motivational systems, of which the behavioral approach versus avoidance system (BAS vs. BIS), as well as the fight-flight-freeze (FFS), are the most widely discussed systems. Extraversion and Neuroticism are the two traits that were central to the original theory. Extraversion is clearly activated by the behavioral approach system, whereas Neuroticism involves both the behavioral avoidance system and the FFS. Openness to change is most likely activated by the behavioral approach system (as well as the pleasure system, which is to date less well described), whereas both Agreeableness and Conscientiousness are activated by both systems. The two neurotransmitters, dopamine and serotonin, are proposed to be most central for regulating the underlying brain systems, but it is evident that a host of other neuropeptides are also involved (even though their roles are not that well understood yet). This theoretical account has been further developed by DeYoung, as described next.

Cybernetic Big Five Theory

Colin DeYoung (2014) built on reinforcement sensitivity theory to develop the cybernetic Big Five theory (CB5T). It incorporates the two meta-traits (stability vs. plasticity) above the Big Five, as well as trait aspects (e.g., industriousness, orderliness within Conscientiousness, and compassion and politeness within Agreeableness) below the Big Five that I discussed in Chapters 3 and 4. Cybernetic feedback loops exist at each of these levels. For example, at the meta-trait level, stability has the function of protecting the system from disruptions by impulses, whereas the plasticity meta-trait is focused on exploration, identifying possible desirable new end-states, and acquiring resources to attain those.

A second and more important extension by DeYoung is the more explicit discussion of the process that leads to the emergence of personality traits and how these in turn shape life outcomes and characteristic adaptations. The main thrust of his argument is that genes and environment jointly influence the states of the cybernetic system, which can be described in terms of relatively stable parameters. A person might be typically higher or lower in a particular motivation (e.g., one person is more outgoing compared to another person), yet at any given moment, there are temporary changes (which may occasionally make the shy person more outgoing when interacting with close friends in a safe environment than the habitually more outgoing person, who at that moment is sick or tired or occupied with other goals). These cybernetic mechanisms are based on incentives and feedback processes that indicate how the system is progressing toward attaining rewards or avoiding punishments. These processes typically unfold sequentially and require (a) goal activation, (b) action selection, (c) action, (d) outcome interpretation, and (e) goal comparison. Therefore, he adds more steps to the basic cybernetic model shown in Figure 6.1. This sequence is obviously schematic, and multiple processes might be operating simultaneously (e.g., sensory information from our eyes, ears, and skin are streaming in and being processed constantly). Limitations in the ability to pursue multiple actions at once will nevertheless restrict the system to a more or less linear sequential state most of the time. While writing this chapter, I am unable to simultaneously pursue my social goals of being with my friends or satisfying my hunger. This cybernetic system relies on both complex planning and information-processing mechanisms, as well as memory systems of goals, actions, and knowledge of the world. These processes were not explicitly part of the reinforcement sensitivity theory and form a novel part of DeYoung's theory. He calls these processes "characteristic adaptations" (in line with the processes discussed by McAdams & Pals, 2006; McCrae & Costa, 1999).

DeYoung defines characteristic adaptations as "relatively stable goals, interpretations, and strategies specified in relation to an individual's particular life circumstances" (2014, p. 33). The most important adaptations are goals (broad representations of desired future states), interpretations (internal representations of oneself and the environment, based on both factual and evaluative information), and strategies (plans, actions, skills, and routines that are used to transform the current state into the desired future state). Distinguishing adaptations from traits, DeYoung argues that traits are culturally universal, whereas adaptations are culturally variable. As we saw in the previous chapter, this is not likely the case in the case of values. DeYoung does acknowledge that values as defined and discussed in Chapter 3 might be seen as personality traits: "many values can be categorized as facets of the Big Five, and these categorizations should be mapped to aid in integrating theories of values with other theories of personality" (p. 40).

The elements that make up characteristic adaptations in this model are central features of a different class of theories of personality. These theories were developed more from the situation side of the person-situation debate. Let's have a look at these theories.

SOCIAL-COGNITIVE THEORIES OF PERSONALITY

The broad set of social cognitive theories of personality start from the assumption that, to understand what a person is like, it is important to first understand how he or she reacts to situations in broad *if* . . . *then* relations (Mischel & Shoda, 1995). These if . . . then relations can then be generalized in terms of prototypical reactions to classes of situations. Therefore, these theories focus more on the situational features. The most widely cited theory – the cognitive affective system theory – was developed by Mischel and Shoda.

Cognitive Affective System Theory

Mischel and Shoda (1995) developed the cognitive-affective personality system (CAPS) theory to explain how psychological mediating processes underlie individual differences (building on earlier work by Mischel, 1973, 1977). They proposed five types of relatively stable person variables as key for understanding personality processes:

(1) Encodings (constructions of the self, people, events, and internal/external situations);
(2) Expectancies and beliefs (about the potential outcomes of a course of action in a specific situation, about self-efficacy, and about the external world);

(3) Affects (emotional and affective reactions, including physiological responses);
(4) Goals and values (desirable/aversive outcomes and associated affective states; goals, values, and life projects); and
(5) Competencies and self-regulatory plans (scripts, plans, and strategies for initiating, maintaining, and modifying actions to obtain desirable outcomes/avoid undesirable outcomes and regulate internal states).

As you can probably recognize, these variables are quite similar to what DeYoung incorporated into his cybernetic Big Five theory. However, Mischel and Shoda did not assume that there are sequential processes (compare this with the sequential steps of cybernetic Big Five theory that I described earlier). Rather, they thought that these variables are organized into interacting units that form the core of personality structure. The availability and accessibility of these units, as well as the patterning of relationships among them in a network structure, are what describe personality. The constellation and interaction of sets of cognitions and affects influence which specific features of a situation are recognized and encoded, which behaviors are selected as appropriate and become executed, and how and which changes in the specific features of the situation are recognized, and so on. There are many possible connections between individual units, but only some connections will be important in a specific situation and therefore will become activated in a specific individual. Feedback mechanisms produce and sustain activations over time. The whole of the activated networks and their distinct organization within an individual ultimately lead to observable behavior as well as distinct differences observable between individuals. In a way, the CAPS is a much more complex and complicated cybernetic system (see Figure 6.2).

For our purposes, it is noteworthy that values are explicitly included in the list of cognitive-affective units that make up the person. Together with goals, values are conceptualized as outcomes that are either desirable or aversive and that motivate the system to act. Although Gray's work was not cited, this recognition of overall goals that a system is trying to achieve or avoid does sound very similar to the broad motivational systems in reinforcement sensitivity theory. Hence, there are similarities in some core features of both sets of theories, even though these theories approach personality from opposite perspectives.

The crucial factor for understanding differences between individuals is the existence of idiosyncratic *if . . . then* relationships. For example, person A might become irritated if she believes that she is being ignored, whereas person B might feel happy when he is left alone and in contrast becomes

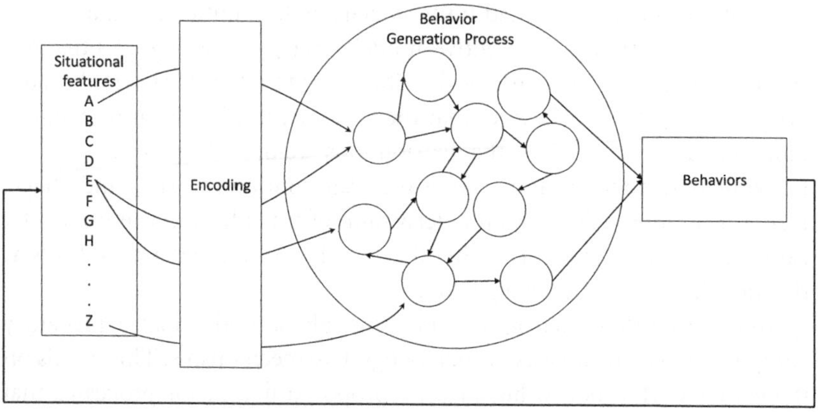

FIGURE 6.2 A schematic model of the network structure in Mischel and Shoda's cognitive affective personality system.

irritated if other individuals intrude on his perceived personal space or share personal stories. Individual differences become understandable when the specific associations and situational meanings are decoded. There might be an overall similar level of irritability, but for each individual, it is driven by distinct and predictable patterns of behavioral activation across classes of situations.

In this theoretical approach, dispositions are thought to generate distinct processing dynamics that generalize to processing dispositions. Processing dispositions include processing structures and processing dynamics. Processing structures are the organized set of cognitions, affects, and behavioral strategies within an individual that guide and constrain activation, and processing dynamics are the patterns and sequences of the activation of mediating units when individuals encounter a specific situation with specific features (real or imagined). These dispositions are highly variable, and Mischel and Shoda made no attempt to link them to a general descriptive theory such as the Big Five. In fact, little phenotypic organization was presupposed, but they did explicitly acknowledge the role of neurocognitive processes (Mischel & Shoda, 1995, p. 260). Their theory inspired other work that attempts to integrate such complex interactions with the Big Five. I discuss one such recent model next.

Whole Trait Theory

To account for the emergence of the Big Five traits, Fleeson (2001; Fleeson & Jayawickreme, 2015) attempted to merge social-cognitive models with trait descriptions. First, he defined traits as density distributions of states. This is consistent with social-cognitive models that propose highly situation-specific reactions. There are wide distributions of trait-like behavior because people change their behaviors that they are manifesting from moment to moment. At the same time, the distributions are thought to differ between individuals in their location along the trait continuum, size, and shape. Hence, this approach allows the derivation of traits from more frequently exhibited states and vice versa. In other words, we can understand traits as the most frequent states of a person.

To describe these density distributions, Fleeson added an explanatory component, which consists of social-cognitive mechanisms. This builds on the earlier CAPS theory, but Fleeson proposed five major processes that differ from Mischel and Shoda's theory and have a much stronger behavioral component (which is similar to reinforcement sensitivity theory, even though it is not explicitly mentioned):

(1) Interpretive processes (information processing at the input level);
(2) Motivational processes (representations of desired and feared end-states that form the incentives for behavior, which seem very similar to the core motivations proposed by Gray);
(3) Stability-inducing processes (any genetic, homeostatic, or habitual forces that guide an individual to his or her trait manifestations; this is a novel aspect of this model compared to CAPS);
(4) Temporal processes (any processes that account for influences of past events, including inertia and cycles; this is similar to the memory process included by Mischel and Shoda); and
(5) Random error processes (necessary to account for unpredictable trait manifestations; this is also a novel aspect of this theory).

Fleeson and Jayawickreme (2015) argue that, in contrast to Mischel and Shoda's work, their model is grounded in the Big Five: their proposed explanatory social-cognitive mechanisms result in the manifestation of the Big Five traits over time. Narrow traits become layered and linked together to form the trait domains. Drawing on network models of personality (Cramer et al., 2012), Fleeson and Jayawickreme argue that individuals learn how to behave appropriately in specific situations. For example, narrow traits of boldness and assertiveness both involve dominating others, with the outcome that a person achieves a certain desired goal if either one of the traits is used. Over time, people learn that either strategy works and start using them interchangeably, depending on the situation. These accretion processes are constrained by biology, culture, logic, and physical realities

(p. 88), resulting in a common structure across a wide variety of individuals and cultures.

As we have seen, social-cognitive theories focus on specific cognitive mechanisms that may give rise to differences between individuals, while specifically addressing the features of the situation and how they might be perceived and processed by different individuals. As they continue to be refined, the most recent models proposed by DeYoung (2014) and Fleeson and Jayawickreme (2015) are becoming conceptually closer, even though there remain major differences between them. However, it is becoming increasingly evident that any adequate account of personality needs to take into consideration both stable predispositions related to broad motivational systems and situation-specific cognitive processes that give rise to behavioral reactions. Neuronal models of personality bridge both by integrating core ideas from both perspectives.

INTEGRATING NEUROBIOLOGY AND SOCIAL COGNITION: NEURAL NETWORK MODELS OF PERSONALITY

The previous sections reviewed two major theoretical approaches and a number of prototypical theories within each group. Both approaches are necessary to understand personality at the deeper level. Neural network models of personality bridge the neurobiological and social-cognitive divide. These models of personality (Read et al., 2010) begin with the postulation of two major motivational systems: an approach system and an avoidance system (derived from reinforcement sensitivity theory). In Read and colleagues' version, the assumption is that the approach motivation is dopaminergic and that avoidance motivation is serotonergic innervated. These broad motivational systems are then linked to motives that are based on evolutionary salient goals that humans need to achieve to survive and reproduce, such as exploring the environment to find resources; gaining higher social status to increase access to resources, forming coalitions to ward off competitors and getting easier access to food and other resources; forming affectionate relationships, finding a partner to mate; and then taking care of one's offspring. These motives can be distinguished in terms of agentic (dominance, assertiveness, etc.) and communal/affiliative motives (affectionate and warm interpersonal relationships).

The various systems and elements (behavioral motivations and evolutionary motives and goals) are hierarchically organized. The motives based on evolutionary goals are situated at the lowest level. The two broad motivational systems of approach and avoidance sit on top of these, influencing and integrating these lower motives. The approach system integrates and moderates motives related to social bonding, dominance and authority relations, exploration and play, and caring for offspring and mating. The avoidance system regulates motives related to avoiding physical harm and social separation. Finally, the authors propose a general disinhibition and constraint system that is orthogonal to the other two systems and moderates all aspects of this intrapersonal model. Read and colleagues argue that this disinhibition and constraint system is associated with serotonergic processes in the brain. In their view, this system serves as a general control system, regulating both the approach and avoidance systems, as well as the lower-level motivational system.

Two other critical components of the model are its representations of the situation and resources. Situations vary in terms of whether they have salient motive-relevant attributes. For example, is a situation perceived to help me gain points with my mates to increase my social status? Or, is this situation helping me get to know that very attractive person whom I would like to talk to? Resources represent the motive-relevant resources that an individual possesses for pursuing specific motives. A person might perceive that a particular situation could increase his or her social status, but it may require some resources to realize its potential. For example, solving a particularly tricky work problem might increase the social status of a worker among colleagues. But to be able to solve the problem, the worker will need to read up on it and gain more knowledge. Without adequate resources, a particular behavior might not be manifested.

Therefore, this model brings together most elements of the various theories that I have discussed so far and integrates them in a complex interplay among motivations, motives, situational features, and personal resources. Motives are activated as a function of situational features, experience of the person (e.g., memory, knowledge), and innate interindividual differences representing the general sensitivity of the approach versus avoidance system combined with the regulatory strength of the disinhibition and constraint system. The expression of motives across situations in turn leads to behavioral regularities that can be described in the form of the Big Five.

To test this model, the authors developed a neural simulation model. Important features of this simulation included the ability of the system to learn (changes in parameter weights within the model), complex mapping of motives and situations (more than one goal could be activated in one situation; goals could be applicable across different situational features), and the possibility of equifinality (different behaviors can lead to the same result) and multifinality (the same behavior can achieve different goals). The network simulation included all these components in a network of nodes that can learn and therefore update their node functions (taking a central idea from Mischel and Shoda's work, these nodes are the small circles represented in the center of Figure 6.2 as the behavior-generating processes). A moderately large number of situational features (at home, in a conference room, at a party, etc.), resources (social skills, job skills, wit, money, things to talk about, etc.), goals (obtaining friendship, sex/romance, exploring, material gain, avoiding guilt, avoiding loss of control, etc.), and behaviors (eat and drink, stay at the periphery of social situations, help others with work, steal, mediate, ignore others, surf the web, etc.) were fed into the simulation, and the parameters were manipulated across different simulation runs to test the importance of the individual system components in shaping behavior choices.

The results suggested that the system could recover prototypical behaviors associated with Extraversion, Neuroticism, and Conscientiousness across a number of situations. Extraversion- and Neuroticism-type behaviors were accurately modeled by modifying states of the approach versus avoidance systems. This suggests that Extraversion and Neuroticism are relatively straightforward in their motivational orientation.

As noted earlier in the section on reinforcement sensitivity theory, Conscientiousness is more complex. Read and colleagues were able to simulate Conscientiousness traits through modifying the overall disinhibition and constraint system. This is a major contribution because it provides a conceptual model of how Conscientiousness as a personality trait may emerge, showing that it is related to behavioral avoidance motivations that monitor potential conflicts by inhibiting spontaneous impulses and favoring the deliberate exploitation of situations.

The simulation was also able to account for the density distributions described by Fleeson (2001). Situational features had a significant effect on behavior (leading to large intraindividual variability across situations), often following *if . . . then* contingencies as proposed by Mischel and Shoda, (1995). Therefore, there were both similarities in behavior expressed by individuals with different dispositions when the situation did not provide incentives to activate motives (e.g., both outgoing and shy individuals will show similar behavior when working alone on a project) but also profound differences in behaviors in a situation that activates a salient motive (e.g., in a social setting with unfamiliar others, outgoing people will approach these unfamiliar individuals, whereas shy individuals are unlikely to initiate contact). Variability within systems across situations was often equal to variability across systems. At the same time, stable interindividual differences emerged in the distribution of prototypical behaviors that reflected the baseline activation of the systems. Overall, the simulation was successful in demonstrating that both specific trait facets and trait domains can be computed based on a small set of motives in interaction with the environment and the ability to learn.

In summary, it is possible to combine core insights from the two general theoretical processes and integrate them in a single model. A simulation combining broad behavioral approach versus avoidance systems with an explicit modeling of situational features, resources, and learning that are central to social-cognitive theories demonstrated that three of the Big Five traits can be modeled in artificial systems. This is exciting and convincing evidence that it is possible to recover such complex traits as the Big Five through a smaller number of basic biological and cognitive systems. However, what plausible mechanisms can explain prosocial traits such as Agreeableness? I take on this challenge in the next section.

PROSOCIALITY AND ALTRUISM IN THE BRAIN

What should be clear by now is that theoretical accounts have accounted relatively well for three of the Big Five. But what about the other two personality traits? The literature suggests that Openness might be influenced by a subset of behavioral approach motivations (specifically focused on exploring situations where new information might be gained). Thus, Agreeableness is the trait that has been least well explained. At the same time, Agreeableness is closely related to values. Self-Transcendence, the most closely related value to Agreeableness, is conceptually similar to the concepts of altruism and prosociality, which have been of major interest to evolutionary biologists (for a general review of the psychological research, see Keltner, Kogan, Piff, & Saturn, 2014).

In biology, one of the core assumptions is that altruism is selfish in that it is typically geared toward people who are genetically related to us. Apparently selfless acts geared toward genetically related individuals are likely to increase the chances that some part of our genes will survive; hence, the term "inclusive fitness." This "selfish gene" hypothesis has been most popularized by Dawkins's (1976) famous book of the same title. But obviously, even unrelated individuals cooperate with each other. To account for these acts among people (and animals) who are not genetically related, Trivers (1971) proposed reciprocal altruism as an alternative mechanism.

The assumption is that initially costly interactions for an actor that are beneficial for a recipient are selected by evolution if they occur over extended periods of time and become reciprocated (the beneficiary of the initial selfless act will reciprocate on a future occasion). It is a theoretically plausible argument that can account for a number of human characteristics, including friendship (adaptive because it motivates us to find and associate with others with whom we can trade altruistic acts), costly punishment (punishing cheaters at one's own cost to uphold moral and normative exchange systems) and gratitude (adaptive because it makes the provider of an altruistic act believe that the recipient of the act is likely to reciprocate on a future occasion; Laland & Brown, 2011).

How do such processes play out in the brain? Integrating a large literature across disciplines, Declerck, Boone, and Emonds (2013) proposed a neuroeconomic model to explain how costly prosocial decisions are computed. Their main distinction was between two situational features that activate two separate brain pathways that jointly lead to a decision to either cooperate or defect. In economics, any situation that requires social cooperation can be categorized as a social dilemma. Thus, in any interaction, people will have incentives for cooperation versus defection, and the incentives' relative strength will determine whether partners will cooperate or defect. There are two major types of incentives. Extrinsic incentives entice people to cooperate because they align the self-interests of the target person with those of the partner or collective. This type of incentive takes various forms, including the perceived benefits of long-term cooperation (e.g., mutual reciprocity or tit-for-tat), synergy (e.g., in a team, people have complementary skills that are all necessary to get a job done; therefore, cooperation is beneficial for all partners), reputation (e.g., increased cooperation and generosity are often associated with increased social status, which brings additional benefits in the future), and the threat of punishment (e.g., it is more costly not to cooperate because of the resulting costs, which may include poor reputation, loss of status, social exclusion, or monetary fines).

In contrast, intrinsic incentives are related to trust signals and a "fear of betrayal." Humans are hypersocial animals, and cooperating with others has become intrinsically motivating. Trust signals within an exchange situation subjectively transform the fear of betrayal into positive expectations of reciprocity and can be intrinsically rewarding. Importantly, trust signals do not require extrinsic incentives to still activate cooperative behavior, presumably because they are intrinsically rewarding. Therefore, extrinsic incentives and trust signals appear to be independent drivers of cooperation.

Declerck and colleagues argued that these incentive situations activate either a cognitive control system or a social cognition system. The cognitive control system is located in the lateral prefrontal cortex (PFC; in particular, the dorsolateral PFC and the lateral orbito PFC) and the dorsal ACC. It has been shown to be activated by calculating tradeoffs and extrinsic incentives. The social cognition system, which processes socioemotional signals and monitors interpersonal trust signals, is located in the medial PFC, the temporoparietal junction (TPJ, including the posterior superior temporal sulcus), and the amygdala. Both systems then interact with the brain reward system that generates the motivation to cooperate (or not). The reward system is thought be modulated by dopamine pathways located in the ventromedial PFC and parts of the limbic system and striatum (the ventral striatum, dorsal striatum, and nucleus caudatum), by opioids, and by other neurochemicals such as oxytocin and vasopressin. Oxytocin and vasopressin in particular have generated much interest in the quest to understand how prosocial behavior might be modulated at a neural level (Ebstein, Knafo, Mankuta, & Chew, 2012; Feldman, Monakhov, Pratt, & Ebstein, 2016). The overall theoretical system can explain current economic and neuroscience evidence reasonably well, suggesting it is a good working model of how cooperative decisions play out in the brain, both for specific situations and for more stable traits and predispositions between individuals.

More importantly, this neuroeconomic model can also account for costly punishment (also called altruistic punishment); that is, situations where individuals are willing to pay a cost for punishing individuals who transgress norms. These behaviors are crucial for the functioning of social institutions because they require an absence of free-riding, which can be achieved by the punishment of those who do violate norms. Altruistic punishment is most likely driven by mechanisms regulating negative emotions (reducing arousal and restoring positive emotionality, an emotion-regulation strategy) and restoring fairness (a cognitive control strategy).

Individual differences in the tendency to cooperate with others can be integrated into this system. Again, individual differences are not sufficient to cause any behavior, but require contextual signals to activate relevant brain processes that then lead to goal-directed behavior. When there are general social preferences to defect (unwillingness to cooperate) stronger extrinsic incentives are needed to induce cooperation. The cognitive control system will then provide the motivation to cooperate. In contrast, individuals who are other-regarding and have general expectations of others are more likely to monitor situations for signals of trustworthiness. These signals are processed via the social cognition system (possibly regulated via oxytocin and vasopressin) and lead to cooperative behavior.

In summary, this model proposes two distinct processes that are associated with cooperation and may be relevant for personality. Agreeableness and Self-Transcendence are clearly associated with a trust-based process, in which individuals cooperate and engage in prosocial acts without expecting anything in return. The trust-based cooperation system is a reasonable neurocognitive mechanism that can explain individual differences (Declerck et al., 2013). The extrinsic system is likely to be associated with Self-Enhancement values, and the major motivation is to increase one's social status and to achieve positive outcomes for one's own advantage.

AN INTEGRATIVE PROPOSAL

As can be seen in this brief review of a rich and diverse literature, personality processes can only be explained if both neurobiological and social variables are attended to. Whatever theoretical processes are being put forward, they should at least be plausible from a neuroscience perspective because all behavior is initiated by brain mechanisms. But at the same time, we should not fall into the trap of neuro-reductionism, because our understanding of how complex psychological processes play out in each individual's brain is still fairly limited (and given classic principles of uncertainty in the natural sciences, we may never be able to precisely state the position and momentum of chemical particles that in their concerted entirety give rise to psychological realities). As Yarkoni (2015) has pointed out quite sarcastically, neurobiological systems do not care about the Big Five.

Modern versions of theories on both sides of the person versus situation debate agree, at least implicitly, that neurobiological systems of approach and avoidance motivation provide an important regulating mechanism for traits, in particular, Neuroticism and Extraversion. A second important dimension that has emerged from evolutionary, economic, and neurocognitive research is the extent to which human cooperation is driven by either trust or instrumental control mechanisms. Humans are a hypersocial species. To survive as a species, we had to cooperate with each other, because our bodies are too weak to compete with other predators and obtain sufficient food resources to survive. We cannot but cooperate. But we need to make sure that we are not exploited and taken advantage of in social relationships. If we cooperate always and unconditionally, we would be highly vulnerable to free-riders (see Dawkins, 1976). Therefore, we humans need to constantly make decisions whether to cooperate or not. Given the centrality that social cooperation has had for us over evolutionary history up until the present day, interindividual differences of how these decisions are handled are likely to play a major role in personality. For this reason, I argue that this trust-based versus instrumental control-based cooperation distinction is the second major dimension that underlies individual differences of how people behave and interact.

I propose that these two processes – behavioral approach versus avoidance motivation and cooperation systems – can explain the spatial organization of the value structure and personality traits. These are not dimensions in the classic sense, but rather are separate systems that interactively modulate motivations and behavior. The basic conceptual model shown in Figure 6.2 is a good start, but certainly will need updating and revisions as more research becomes available. I (arbitrarily) place values in the inner circle. To recap, values are subjective evaluations of what is important for a person at a given moment; they are desirable goals and standards. Traits, in contrast, are typical behaviors and dispositional tendencies. Traits and values are not identical because they involve different neurocognitive processes (values have a much stronger evaluative component). Yet, both values and traits are influenced by the same underlying genetic and neurocognitive processes, which is why I combined them in this figure. In other words, values and traits are instantiations of the same major neurobiological processes and are regulated and organized by the same biological systems operating within our brains.

In the rest of this section, I briefly discuss how the findings from the previous chapters fit with the integrated behavioral approach vs. avoidance motivation and cooperation systems proposal (see Figure 6.3). To simplify the arguments, I use the higher-order values to organize the discussion.

The trait of Openness is primarily associated with Openness to change versus Conservatism values. Openness to change values are focused on approaching new and novel situations; therefore, it is a straightforward prediction that Openness values and Openness trait are under the control of the BAS, potentially linked to a more specific "liking" component driven by opioid systems (and potentially innervated by a separate Pleasure System).

Extraversion is also related to Openness to change versus Conservatism values, but has some secondary correlations with Self-Enhancement values. To some extent, Self-Enhancement values also involve motivations to engage with new situations and take advantage of opportunities, which require substantial activation of approach motivations. Neurobiological theories reviewed earlier provide strong empirical support for the

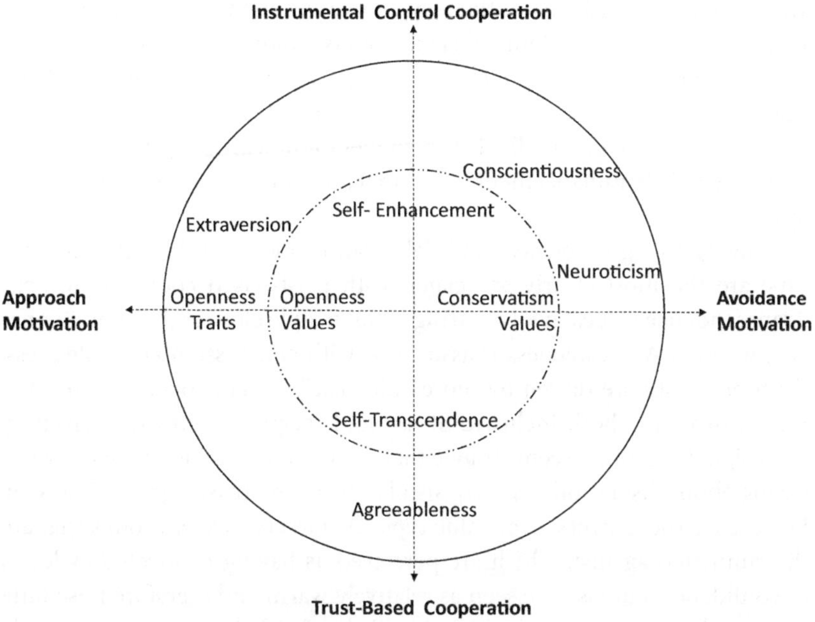

FIGURE 6.3 The overall framework of values and traits based on the integration of behavioral approach versus avoidance motivation and forms of cooperation systems.

proposal that Extraversion is regulated by the BAS. A prediction worth testing that derives from this organization of traits and values is that Openness to change values are particularly strongly related to the enthusiasm trait aspect of Extraversion.

Extraversion and Self-Enhancement values require cooperation and social exchanges. These behaviors are driven by instrumental control considerations, because their focus is on how to gain status and social approval. This should be particularly relevant for the assertiveness aspect of Extraversion. Such a close association between assertiveness and Self-Enhancement is testable in future studies. The correlation between Extraversion (especially the assertiveness aspect) and Self-Enhancement described in Chapter 4 is consistent with a proposed underlying activation of instrumental control cooperation mechanisms.

Moving around the circle, Conscientiousness is correlated with both Self-Enhancement and Conservatism values. As simulations by Read et al. (2010) have demonstrated, Conscientiousness is strongly influenced by overall inhibition and constraints. Inhibiting impulses that may upset others or the traditional social order is another key motive within the Conservatism value cluster. Therefore it is highly likely that conservative values and Conscientiousness are both modulated by the same inhibitory processes.[1] Yet Conscientiousness also often requires engaging with others to get projects done. This requires instrumentally driven cooperation, which explains the correlations with Self-Enhancement values that are also most likely driven by this system.

Moving to Agreeableness and Self-Transcendence, this trait and value type are the most clearly associated with trust-based cooperation. Self-Transcendence values involve caring about people close and distant to oneself, whereas Agreeableness is associated with compassion and politeness. Both processes are driven by more "intrinsic" social cooperation mechanisms, involving both inclusive fitness and reciprocal altruism driven by trust signals. It might seem strange that Universalism values, which involve caring about distant others, may also be driven by trust signals. That may be because the Universalism value typically targets weaker groups that are discriminated against, which are perceived as having relatively low levels of confidence, but also are seen as relatively warm and therefore pose little threat (which is a trust signal; Cuddy, FIske, & Glick., 2007, 2008; Cuddy et al., 2009). In addition, reciprocal altruism signals have been exploited by cultural evolution processes in the creation of institutions that effectively use and manipulate trust signals (e.g., in-group markers that signal trustworthiness).

Agreeableness and Self-Transcendence values are partially innervated by both the BIS and BAS, being processed in largely overlapping brain areas (e.g., Brosch et al., 2011). They are approach motivated because they involve engaging with others in social situations, but also require some monitoring of impulses or goals that may potentially upset others. Hence, their proposed position is in the middle between the two extreme ends of the BIS and BAS.

I skipped Neuroticism just now because this trait does correlate relatively weakly with values at a phenotypical level. At the same time, these correlations are systematic and appear to be quite strongly driven by the same underlying genetic structure, indicating that values and Neuroticism might be influenced via similar neurobiological systems. This highlights one important difference between trait personality approaches and value

[1] There is also evidence that the processing of Conservatism values is related to brain areas that are involved in error monitoring (Brosch, Coppin, Scherer, Schwartz, & Sander, 2011; Brosch & Sander, 2013). This is a central area of the BIS, suggesting that Conservatism values are processed in similar brain regions as personality traits related to behavioral inhibition.

research that I highlighted earlier. Trait inventories list behavioral traits or states (I worry, I feel blue, I am often angry, etc.), but these statements do not indicate the relevant goal of the behavior. In contrast, values capture what is important in the minds of a person: they are abstract goals (Schwartz, 1992; Schwartz et al., 2012). If we were to reconceptualize Neuroticism as goal-related behaviors, probably more consistent correlations with values would appear. I predict that Neuroticism-related goals broadly correlate with values in the Conservatism cluster and potentially with some Self-Enhancing values. For example, withdrawal might be motivated by goals not to upset others and the traditional order. If the rules and norms are unclear and avoidance goals are activated, the safest strategy would be to withdraw from situations. This may also explain the positioning of orderliness as a Conscientiousness aspect relatively close to the withdrawal component that we observed in the MDS analysis of the DeYoung et al. (2007) data in Chapter 4 (see also Read et al., 2010). In contrast, the volatility aspect of Neuroticism might be motivated by emotionally driven attempts to dominate others to increase one's self-esteem or react to perceived threats. These proposals can be tested with research examining personality traits that include explanatory goals for specific dispositional tendencies.

In summary, I propose that values and personality traits are related to each other by the joint regulation of the same underlying neurobiological systems. The most likely systems are an approach versus avoidance motivation system and an instrumental control versus trust-based cooperation system. At the same time, as the various neurobiological, social-cognitive, and neural network theories have shown, to understand behavior we have to also consider situational features and resources. This issue returns to center stage in the next chapters.

SUMMARY

This chapter provided a tour de force review of various models proposed to explain the underlying reasons for the Big Five structure. These models do not always agree on the details, but if their core features are integrated, they do a good job in explaining the Big Five. I also drew on altruism research to explain possible mechanisms that underlie both Agreeableness and the Self-Enhancement versus Self-Transcendence value dimension. When we put the motivational and cooperation dimensions together, they provide a reasonable account of the empirical associations that we had found in Chapter 4. So as not to overwhelm you with too much information, I had to gloss over some of the finer details of the brain mechanisms. For example,

I highlighted the role of serotonin and dopamine, keeping in line with the established literature. Yet, other neurotransmitters and neuropeptides are likely to be involved. Exploring the role of the integrated set of neurotransmitters is an exciting area for future research (I return to some of these ideas in the final chapter).

Before moving on, let's return briefly to Tina and Sandra. We noted their very different behaviors, but the previous chapter had suggested that their underlying personality traits and values might be more similar than we would expect. The current chapter now provides a snapshot of the underlying mechanisms that can explain these similarities. Their personalities are likely to be driven by a smaller set of basic biological systems that we humans all inherited from our ancestors. But these biological systems interact with the environment that Tina and Sandra are in and the different resources that are available to them. The next chapter takes a closer look at the role of the situation and resources in organizing the structure of personality.

7

Is the Personality World Two-Dimensional?: The (in)Stability of the Trait-Value Structure across Cultures

So far we have worked on the assumption that there is a coherent structure organizing values and personality, which can be recovered in samples from around the world and is neatly organized in a multidimensional structure with two major axes. Tina and Sandra, whom we met in Chapter 1, could be described using the same dimension as a randomly chosen person working in a supermarket in rural France or the manager of a new start-up in Lagos, Nigeria. To what extent is this two-dimensional structure applicable to all humans? To address this question, it is necessary to introduce a fundamental concepts in survey research relating to the equivalence of data. Equivalence deals with the inferred meaning of a data point, answering the question whether we can trust the score that we got from our survey. Unfortunately, it is not an easy question to answer. In this chapter I present some data showing that the two-dimensional world of personality (or the five major dimensions of the Big Five) is not so smooth as is often assumed and conveyed in textbooks. The data often fit reality much less well than we would like, and I lay out the messy evidence here. In the next chapter, I address whether there are some systematic patterns that can explain why the theoretical organization of personality works better (or worse) in some parts of the globe compared to others. Hold on tight, because this will be even more of a geeky chapter than earlier ones, but we need the information it presents to make sense of what is going on.

MEASUREMENT INVARIANCE IN CROSS-CULTURAL COMPARISON

Personality traits or value orientations are latent constructs, which means that they are not directly observable and need to be inferred through measurement indicators that provide insights into the levels of the assumed underlying trait in each individual. Once we have the data point, we then make the inverse reference: a latent variable (that is not directly observable) is assumed to cause the behavioral expressions that we observed. Typically, the responses to items asking about the importance or frequency of particular behaviors or what is important in a person's life are taken as behavioral expressions that provide insights into the underlying traits or values. I strongly believe that the latent variable approach is a convenient starting point for research, but it raises some significant theoretical and philosophical challenges that are considerably beyond the scope of what I can discuss here (see Borsboom, 2006, 2008; Borsboom, Mellenbergh, & van Heerden, 2003). In this chapter I focus on the issues that matter for culture-comparative research.

If we want to measure any psychological construct, we need to ask a few simple questions. First, can the theoretical construct account for (test) behavior in each group? For example, is Neuroticism a plausible theoretical construct to account for a person's responses to items such as "I often worry"? Second, can the same items be used to measure the theoretical variable in each group? Is an item such as "I often feel blue" a good indicator of Neuroticism? It may not be because the names of colors often carry symbolic surplus meanings, which may change the implied personality associations. As an example, in German "blue" is associated with intoxication, whereas in China it might imply immortality. Third, can the responses on a scale be directly compared with each other? This refers to the underlying metric: is a response of 5 on a rating scale that ranges from 1–7 comparable across different groups? To decide whether a mean response of 5 is comparable across groups, we have to examine two different statistical parameters. Therefore, the three major questions (with the last question requiring two subquestions about two different parameters) lead to four levels of measurement invariance. I introduce the major concerns and conditions of each of these four levels. My aim is to provide an accessible introduction to these issues and let readers consult more technical discussions that are available elsewhere (de Vijver & Leung, 1997; Fontaine, 2005; Lun, Fischer, & Ward, 2010; Meredith, 1993; Vandenberg & Lance, 2000).[1] I focus on personality and values and address specific issues and details in Table 7.1.

The weakest or lowest level of measurement invariance is *functional invariance*. It corresponds to the first question posed earlier; namely,

[1] I use the term "measurement invariance" throughout this book. Measurement invariance is used widely in the psychometric literature and is conceptually related to the latent variable approach that guides much of the personality literature. There is a related literature that

TABLE 7.1 *Levels of Invariance Applied to Trait and Values Research*

	What Is Being Tested	What Conditions Need to Be Satisfied	What Does It Mean	What Can We Infer from the Data
Functional invariance	The same underlying trait or value type accounts for responses in the measurement instrument across cultures.	There is a similar network of convergent and discriminant relationships with other constructs across cultural groups.	Respondents identify this trait or value type as meaningful for describing their behavior or motivations.	The trait or value type has broadly similar functions in each culture, supporting general definitions about traits or value types.
Structural/ Configural invariance	Can the same trait/value items be used across cultures?	Items need to be relevant and representative of the underlying trait/value. Items have the same internal structure (a nontrivial relationship with the trait/value) across cultures.	Same items can be used to measure traits or value types, and individuals in different cultures use linguistically equivalent trait and value terms.	
Metric invariance	Can score patterns (trait/ value profiles) be compared across cultures? For example, is A higher or lower than O for an individual and/or sample?	Items have the same weight parameter across cultural groups (each item has the same quantitative relationship/loading with the presumed latent underlying variable).	Respondents associate value items in the same way with the underlying value type. They have the same understanding of the meaning of the item.	Correlations and patterns of means can be compared across cultures. Means are comparable if interpretation stays at the level of the instrument (e.g., the mean of E items in the specific personality test is higher in sample X compared to sample Y, but no claims about an underlying latent trait E can be made).
Full Score/ Scalar invariance	Can we directly compare trait/value scores across cultures? Does a score of 4.3 on a scale from 1–7 have the same meaning in each sample?	There are identical intercepts across cultures. The baseline or intercept (see Figure 7.2) for each item is the same across samples.	Respondents use the response scale in exactly the same way. An expressed score reflects the true underlying motivation of each value type as expressed in the item in the same way.	Scores can be directly compared across cultures, and inferences about the latent underlying latent personality traits and latent values can be made.

Note: I use "cultures" to refer to different samples; the psychometric criteria are derived from Vandenberg and Lance (2000).

whether the latent variable has similar functions across contexts. If a researcher argues that a construct (e.g., Neuroticism or Extraversion) shows functional invariance, then the researcher assumes that the construct serves similar functions in two or more different cultural samples. This means that Extraversion "exists" in each of the cultures and that extroverts in each culture behave generally similar – being outgoing, sociable, and assertive in ways that match local social and cultural requirements and standards. The common method to assess functional invariance empirically is to examine whether the construct of interest shows similar relationships with other related constructs (convergent validity). These relationships can be positive (if the constructs are scored in the same direction) or negative (if the constructs are scored in opposing directions or the construct used for validation captures an opposing construct). For example, Neuroticism should correlate highly and positively with other measures of depression and anxiety, but should correlate negatively with positive mood. To establish functional invariance, a second relationship pattern that should be examined is discriminant validity, which refers to the situation where another construct measured with the same method shows no systematic relationship with the focal construct. For example, we would not expect meaningful relationships between Neuroticism and self-reported creativity or intelligence. Overall, the best approach to examine whether this level of invariance is met is through qualitative research and theoretical reflection that examine relationships between theoretically related and unrelated variables (a nomological net) within each cultural context. Determining functional invariance does not require that the same items be used in each cultural context; it is possible to use different items in each sample. For example, we could include items about spontaneously speaking up in a meeting of village leaders in a traditional Fijian community or addressing a student assembly at a major university event. The important issue for functional invariance is that the construct does exist in the cultural context and produces functionally equivalent behaviors.

There are no statistical tests that can be used to assess functional invariance, because it is primarily a philosophical or qualitative statement about

uses the term "equivalence," but does not assume latent variables. The equivalence literature focuses on the comparability of scores (Poortinga & Vijver, 1987). Despite their different philosophical orientations, they are associated with the same set of statistical tests and conclusions (Fontaine, 2005). Therefore, I draw on both without further differentiating their historical and philosophical legacy. The equivalence literature presupposes an additional level (functional equivalence), and I include it in my treatment here, labeling it "functional invariance."

the theoretical comparability of psychological constructs. We can only evaluate relative relationships; for example, is a particular behavior functionally related to some specific task? Returning to our example of Tina and Sandra from Chapter 1, achieving the important evolutionary task of gaining recognition and social status in one context might involve engaging in religious rituals that show one's credibility, whereas in other contexts it may involve doing hard work as a member on committees. Both behaviors show commitment and dedication that then become socially recognized and elevate the social status of the actor, but through very different behavioral means. In terms of evaluating these behaviors, we could only say that each behavior achieves the particular task.

The next higher level is *structural or configural invariance*. This level of invariance answers the second question proposed earlier: can the same items be used to measure the construct of interest? It typically requires that the same items[2] are used in each sample and that all items are related to the same underlying variable. Conceptually, it means that high levels of Neuroticism "force" an individual to agree with a specific Extraversion item. Statistically, it implies that items load on the same factor or are located in the same geometric space in a multidimensional scaling analysis. The assumption is that the latent variable has some causal relationship with all the items, but the relative strength of this causal relationship may vary across samples. The items "I am the life of the party" and "I stay in the background in social events" both load on the same factor (but in opposing directions), indicating a broad Extraversion trait in two or more samples. The loadings (the extent to which each item taps into the underlying construct) nevertheless may still differ across samples, and therefore, it is not possible to compare the averaged means or correlations across samples.

If functional and structural equivalence are met, the next two levels are important, because they will determine how any empirical data can be interpreted. The next level, *score invariance*, addresses the third question about the empirical comparability of the scores, but we first need to distinguish between two statistical parameters: the relationship between the item and the latent variable (the slopes or loadings) and the ability of the latent variable to predict all the variability in an item (which refers to the intercepts).

[2] It is possible to use only partially overlapping sets of items, with one subset of items being identical in all cultural groups, and a second subset measuring the construct in culture-specific ways in each culture. The statistical treatment of such data is complex, and therefore, it is not often done in the personality and value literature.

These two parameters are associated with *metric invariance* and *full score or scalar invariance.*

The first issue that we have to consider is whether the loadings of items on the latent underlying constructs are identical across cultures. Structural or configural invariance only assumes that there is a nontrivial relationship (an Extraversion item is related to Extraversion without specifying the degree of relatedness). Metric invariance requires that the relationship (the regression coefficient) between an individual item and the underlying latent variable is identical across samples (within the boundaries of statistical probability). This means that an increase in a response (e.g., answering with a 3 on a 1–7 Likert scale instead of a 2) is associated with the same increase in Extraversion (the latent variable that is thought to cause the responses to the survey item) in all samples. There are a number of tests that measure metric invariance. The most typical tests compare the factor loadings of items on the latent factor in a factor analysis (either exploratory or confirmatory factor analysis) or the coordinates of items in a metric MDS. Statistical agreement indices are available to evaluate the statistical similarity in exploratory analyses, such as factor analysis or MDS. Values of 0.90 or higher for the most commonly used agreement indices, such as Tucker's Phi, indicate adequate agreement. In confirmatory analyses (e.g., confirmatory factor analysis, item response theory models), constraints on the similarity in factor loadings can be imposed, and the subsequent deterioration in model fit can be statistically evaluated (which provides much more convincing evidence than other methods). Fischer and Fontaine (2011) discuss how to conduct these analyses in more detail.

What can we conclude if conditions of metric invariance are met? Certainly we are then able to compare correlations and patterns of means (e.g., profiles) directly across cultures. For example, we may conclude that Extraversion is significantly higher than Openness in a New Zealand student sample, but Openness and Extraversion are not different from each other in a Samoan student sample. However, although we can compare instrument means across samples, when metric invariance is met we cannot make claims about any latent underlying construct (see Fontaine, 2005). For example, we could say that the Extraversion scores when measured with the NEO-PI-R (Costa & McCrae, 1992) are significantly higher in sample X compared to sample Y. However, this does not mean that the latent trait of Extraversion is higher in sample X compared to sample Y. Therefore, the interpretation of the meaning of the difference is ambiguous because it has to stay at the level of the particular score and instrument. These differences cannot be taken to indicate any psychological differences in Extraversion

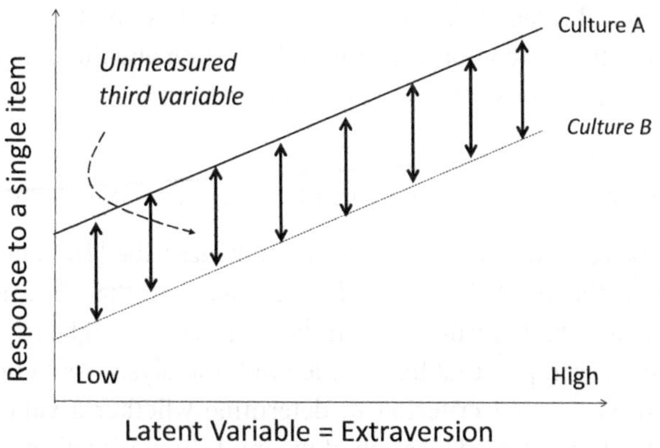

FIGURE 7.1 Example of metric equivalence, where an unknown and unmeasured third variable adds variance to an item that is independent (not "caused") by latent extraversion.

(defined as a latent psychological variable) between samples. This is a subtle but important point.

The reason why we cannot interpret these differences in psychological terms is that the intercepts or baselines may differ between samples. Metric invariance indicates that the slopes between items and latent variables are the same, meaning that an increase in the latent variable is associated with the same increase in the measured item across samples. Yet, there can still be differences in the baselines that are independent of the latent variable (see the intercept differences between the two slopes in Figure 7.1). In sample A, the individuals with the lowest latent scores (the more introverted individuals) still show substantively higher scores on the measured Extraversion item compared to individuals with the same latent score on the underlying Extraversion variable in group B. It suggests that some other (unknown and unmeasured) variable is influencing the score independent of the construct of interest. Extraversion does not explain all the variability in this item, and some other variables are still relevant in explaining the difference between the two samples. In survey research, response sets such as acquiescence are likely candidates for such shifts in baselines (Fischer, Fontaine, van de Vijver, & van Hemert, 2009; Smith & Fischer, 2008; Smith et al., 2016).

The condition where item intercepts are assumed identical across samples is called *scalar* or *full score invariance*; it is the highest of the four levels. This level of equivalence is difficult to meet, and only relatively recently have powerful statistical methods become available for its widespread use in psychological research (Sireci, 2011). Once we have met full score invariance, we can directly compare means and interpret any differences in terms of the assumed underlying psychological construct.[3]

HOW WELL DO VALUES REPLICATE ACROSS CULTURES?

The descriptive theory of values was first fully described by Schwartz in his highly influential article in 1992 (Schwartz, 1992). At that time, it was difficult to evaluate whether values would show sufficiently high levels of invariance across the samples that he included in his analysis. As a consequence, Schwartz used a visual criterion to determine whether a value item was positioned in the expected location of the multidimensional space. He used a criterion of 75% correct placement across all his samples. This suggests that he found evidence of construct invariance. In a more rigorous analysis, Fontaine and colleagues (2008) reanalyzed the original data with more sophisticated methods than were originally available. The team used procrustean target rotation, which helps overcome random fluctuations in the coordinates and makes direct comparison of individual samples straightforward. They then compared the relative fit of items to the theorized two-dimensional structure. Recall that fit can vary between 0 (no fit) to 1 (perfect fit). The reported values varied widely, ranging from 0.12 to 0.53. In a later analysis using more samples, we reanalyzed the structure of values in samples of teachers in 53 nations (Fischer, Vauclair, Fontaine, & Schwartz, 2010). Again, we found considerable variability, ranging from a low of 0.25 in Egypt to a high of 0.70 among a sample of Portuguese teachers. The average similarity was 0.53, indicating that there is considerable variability in the position of individual items across samples.

These analyses used relatively weak statistical methods. Using more statistically powerful methods such as confirmatory factor analysis, Davidov, Schmidt, and Schwartz (2008) found metric equivalence in European samples that used a short version of the Portrait Value Questionnaire (PVQ). However, they needed to free up a number of parameters and to combine several value types before a good fit was found. Confirmatory factor analysis does not easily allow an investigation of the circular structure, so their analysis cannot help us much to determine whether the structural organization was replicated.

3 It is possible to find partial invariance, where invariance is met for most, but not all, items. There remain some conceptual ambiguities, especially if the proportion of noninvariant items increases (see Poortinga & Van de Vijver, 1987).

Focusing on the structure directly, a large scale meta-analysis (Steinmetz, Isidor, & Baeuerle, 2012) of 318 samples (total N = 251,239) from 65 countries found large differences in the structure across the samples. They conducted cluster analysis to identify clusters of countries, instruments, and samples that showed similar structures. They found eight major clusters that were more homogeneous. Four of the clusters showed adequate fit for a circular structure (in particular, clusters consisting of samples from Germanic and Latin countries and those that used the PVQ instead of the SVS). What was striking, however, was that even if the empirically observed structures did not statistically fit the circular structure proposed by Schwartz, all but one of the clusters showed a very clear distinction between Self-Enhancement and Openness to change values on one end and conservative and Self-Transcendence values on the other end. This finding supports the argument made in the previous chapter that approach versus avoidance motivation might be a central distinction that organizes the value structure. This analysis also suggested that Self-Transcendence values might be more strongly associated with avoidance approach motivation (in the previous chapter I speculated that Self-Transcendence involves both approach and avoidance motivation).

In summary, the available evidence suggests that the prototypical value structure developed by Schwartz (1992) describes the value space reasonably well across a large number of samples, but there is a lot of variability in the prototypical organization. Recent applications of more sophisticated methods like multigroup confirmatory factor analysis suggest that metric equivalence can be achieved, but they do not allow us to compare these scores in terms of latent variables. What could account for such cross-cultural variability in value organization? Could that mechanism provide some novel insights into human functioning and psychological processes? Before I try to answer those questions, let me briefly review the evidence for traits across cultures.

HOW WELL DO PERSONALITY TRAITS REPLICATE ACROSS CULTURES?

Evidence Using Big Five Inventories

As we saw in Chapter 2, there is general consensus that personality in Western samples can be described in five major factors: the so-called five factor model (FFM) or the Big Five of personality (Goldberg, 1993; John et al., 2008; McCrae & John, 1992). Applications of various Big Five instruments in

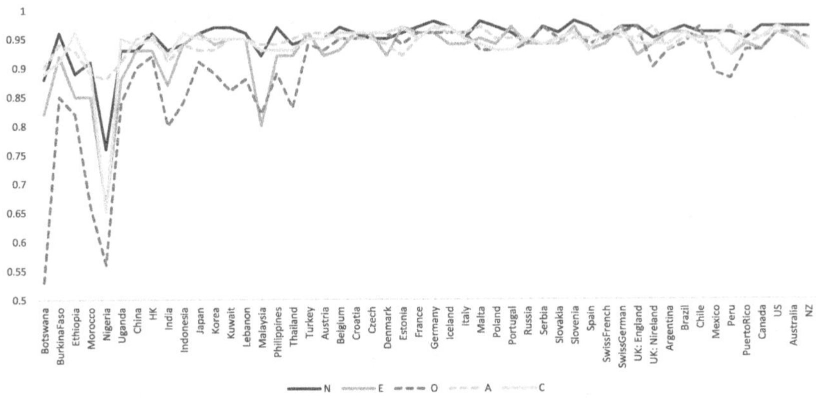

FIGURE 7.2 Congruence coefficients for observer reports of the NEO-PI-R reported in McCrae et al. (2005). *Notes:* A = Agreeableness, C = Conscientiousness, N = Neuroticism, O = Openness, E = Extraversion.

mainly student samples from around the world showed relatively consistent levels of replication in most Western nations, but with poorer replicability in many of the African and some Asian samples (Church et al., 2011; McCrae, 2001; McCrae et al., 2005; Schmitt et al., 2007). For example, McCrae and colleagues (2005) asked students in 50 countries to describe different target individuals using observer rating forms of the NEO-PI-R described in Chapter 2. They conducted their analysis with clusters of items combined instead of using the individual items (which they called trait facets; see Figure 4.6 and the discussion of facets in Chapter 4). The authors found reasonably good agreement between the typical structure found in US samples and most of the specific country samples. The mean congruence coefficients ranged from .90 (Openness) to .96 (Neuroticism). At the same time, there was considerable variability (see Figure 7.2), especially for Openness to experience, but also for Extraversion and Conscientiousness. As mentioned, these analyses were not done with individual items, but with item bundles based on the facets that are organized below the Big Five. Conducting analyses of combined items at the facet level instead of individual raw items typically makes the results considerably more stable and easier to replicate. Finding that much variability when using more stable item bundles is not the best news for the cross-cultural generalizability and validity of the Big Five.

In a different study with student samples from 26 countries, Thalmayer and Saucier (2014) tested a short version of an instrument that measures the six personality trait factors that are compatible with the HEXACO personality model. The final scale showed mixed replicability of the factor

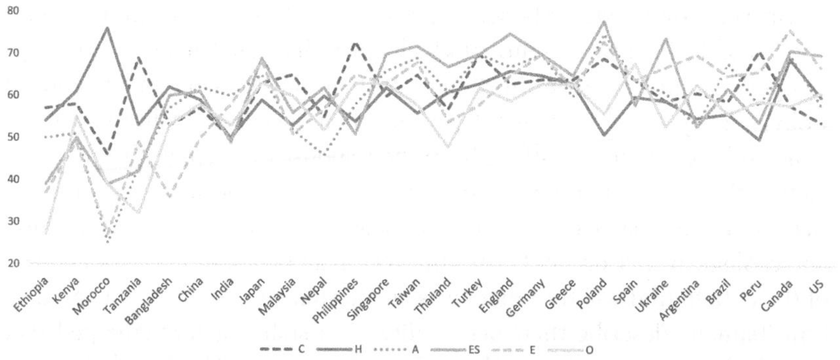

FIGURE 7.3 Reliabilities per sample for each personality trait of the HEXACO in Thalmayer and Saucier (2014). *Notes:* C = Conscientiousness; H = Humility; A = Agreeableness; ES = Emotional Stability; E = Extraversion; O = Openness.

structure. Even models of configural invariance (which imply that all items load on their intended factor) showed relatively low fit. Equality in factor loading and intercepts, indicating metric and scalar invariance, was typically not found. Not surprisingly, the reliabilities varied widely (see Figure 7.3), with particularly large variations in African and Asian samples. Typically, researchers want reliabilities of 0.70 or higher. As can be seen in Figure 7.3, few samples and traits showed sufficient reliability. This analysis was done at the item level and showed greater variability compared to the analysis at the facet level conducted by McCrae and colleagues reported earlier. When analyzing items, there is considerably more noise in the data that might affect the structures compared to the facet-level analyses that are typically reported for the NEO-PI-R. However, this study's findings call into question whether the Big Five traits are reliable and valid across cultures outside North America and Western Europe, at least when using such short scales.

All these studies reported factor analyses, which do not tell us much about the underlying structure. Since correlations are typically not reported between the Big Five, it is difficult to determine whether the relative organization of the factors is replicated across these diverse cultural samples. One important aspect of the framework that I put forward in the previous chapter (see Figure 6.3) is the relative organization of each of the Big Five traits to each other. If the individual correlation matrices per sample were available, we could easily examine whether the relative organization is preserved by reanalyzing the existing data. A more fundamental question, however, is whether the same Big Five factors emerge when we start examining samples not made up of highly educated university students in Western societies.

Gurven, von Rueden, Massenkoff, Kaplan, and Vie (2013) administered a version of the Big Five Inventory (which I introduced in Chapter 3) to a large sample of Tsimane forager-horticulturalists in the Amazonian lowlands in Bolivia. They asked participants to rate their own personality using this scale and also obtained ratings from their spouse or partner. The Tsimane factors showed poor congruence with the US typical structure, suggesting that the Western structure was not applicable in this very different cultural and ecological environment. The authors ran an exploratory factor analysis of the data to find whether there was an alternative structure that captures how Tsimane describe their personality. Two stable factors emerged that were replicated across both self- and spouse reports. The first factor captured a number of Extraversion and Agreeableness items; the researchers labeled it Prosociality. The second factor was defined by high loadings from a number of Conscientiousness items and was called Industriousness. We could criticize this study by arguing that this semiliterate population may not have understood the items or the instructions. Yet, the structure was not random; it was replicated in the spouse data. The authors also reported that splitting the sample into different age or education groups did not change the overall results (individuals received between 0 and 12 years of education, but the overall education level was very low, with respondents attending fewer than 2 years of school). These consistent results do not easily fit with explanations that the items or instructions were not applicable in this population.

Even in industrialized non-Western samples different factor structures might emerge. I administered a short version of the International Personality Item Pool (Donnellan, Oswald, Baird, & Lucas, 2006), containing four additional items measuring Honesty-Humility (Milojev, Osborne, Greaves, Barlow, & Sibley, 2013) to a traditional Buddhist-Taoist community sample in southern Thailand. Both a confirmatory factor analysis and exploratory factor analysis with procrustean rotation showed very poor fit with the proposed structure. Using a separate student sample in a follow-up study again showed that the structure did not fit the typical one found in Western samples. However, four dimensions emerged that were replicable across the student and community samples: two socially oriented factors (one contrasting approach vs. avoidance motivation in the social sphere and the other one capturing empathy-related items), one clear egoism factor, and one factor resembling Neuroticism. What was noteworthy was the strong differentiation of social aspects, suggesting that social elements are an important means by which participants make distinctions when describing themselves.

The Evidence from Lexical Studies and Culture-Specific Instruments

The studies just reviewed applied existing questionnaires to culturally diverse samples. A different option is to start off with person descriptions in each language and to derive a factor structure based on culture- and language-specific data. Trait research is based on the lexical hypothesis that I discussed in Chapter 3 and initially was conducted using English and a few select European language. What happens when we use non-European languages? A number of culture-specific studies have been conducted over the years to answer that question. They all suggest that additional factors or more differentiated factors may emerge within non-Western cultures. For example, Cheung and colleagues (Cheung, Cheung, Howard, & Lim, 2006; Cheung, Fan, & To, 2008; Cheung, Leung, Song, & Zhang, 2001; Cheung, Leung, Zhang, et al., 2001; Cheung, van de Vijver, & Leong, 2011) reported an additional interpersonal relatedness factor when Chinese was used with Chinese samples, whereas a study of person descriptions in 11 indigenous languages in South Africa (Hill et al., 2013; Nel et al., 2012; Valchev, van de Vijver, Rothmann, & Meiring, 2013) found a larger number of clusters that differentiate social-relational aspects of relevance in more collectivistic and group-oriented cultures. In these studies, major factors with close resemblance to the Big Five still emerged, but they were supplemented by additional factors that made finer distinctions among socially relevant behavioral traits. In contrast, in a lexical study of Hindi, a language widely spoken in India, three factors emerged that had very little overlap with any of the Big Five factors commonly described in Western research (Singh, Misra, & Raad, 2013). Instead the three factors were interpretable in relation to ancient Indian philosophy and religious scriptures. One factor captured content that is related to Agreeableness, whereas a second factor resembled a mixture of Conscientiousness traits and Conservatism values. These two seemed to still map to some extent onto the two major axes that I described in the previous chapter (see Figure 6.3). A third factor had loadings from items that were unrelated to any of the Big Five factors, identified by items at one end that included shrewd, asocial, crooked, imposter that contrasted with other items like lively, strict, self-confident, and hospitable at the opposing end. The authors interpreted it as a factor that captures a broad set of behaviors that are functional in a complex and hierarchically structured society such as India (see also Ferreira, Fischer, Porto, Pilati, & Milfont, 2011).

These studies used lexical approaches within single societies. What happens if we extract factors across multiple languages, focusing on the communalities across the various languages and personality dictionaries? Recent studies that simultaneously examined trait dictionaries across a larger number of languages have been able to reliably recover only two (Saucier, Thalmayer, & Bel-Bahar, 2014; Saucier, Thalmayer, Payne, et al., 2014) or three (De Raad et al., 2010) of the Big Five factors. For example, De Raad et al. (2010) reported that Extraversion, Agreeableness, and Conscientiousness formed the most consistent and stable personality configuration across their samples. They also identified the other factors as proposed within the Big Five and the HEXACO model, but when the authors extracted more factors than these three, the individual make-up of the factors started to diverge across this larger number of languages. Saucier, Thalmeyer, and Bel-Bahar (2014) reported a first study across 12 mutually isolated languages and, when using a small set of words that were common to all languages, reported only two common factors. In a subsequent study, Saucier, Thalmeyer, Payne et al. (2014) confirmed these two factors, which they labeled Social Self-Regulation and Dynamism. The former was strongly related to aspects of communion, morality, warmth, interpersonal nurturance, and the absence of externalizing problems. The latter captured agency, competence, dominance, and an absence of internalizing problems. Using the terminology that I introduced in the last chapter, the first seems to capture trust-based cooperation, whereas the second factor appears to be a mix of approach motivation and instrumental cooperation.

De Raad and colleagues (De Raad et al., 2014) combined lexical analyses with statistical techniques to determine which lexical terms go together. Their study therefore blended the techniques used by survey researchers using established inventories with the lexical trait approach that treats trait dictionaries in each language separately. They used 922 trait terms that were found in at least 2 of the 10 languages (and 11 samples) included in their analysis. One limitation was that they only included one Asian language, and the other nine languages were spoken in Europe (as well as English being spoken in a number of other countries around the world). Replicating findings from their earlier study (De Raad et al., 2010), they again found that Extraversion, Agreeableness, and Conscientiousness formed the most stable organization across all the languages. When examining a two-factor structure as indicated by Saucier, Thalmeyer, Payne et al. (2014), Conscientiousness merged with the other two factors, leading to the emergence of two factors that resembled those identified by Saucier and colleagues.

Importantly, both five- and six-factor solutions produced meaningful results, but these factors were not always strongly present in all of the languages. For example, Filipino which is the lingua franca in the Philippines, often showed weaker factors. Interestingly, the global Conscientiousness factor consisted of both Conscientiousness and Emotional Stability (reversed Neuroticism), which resembled some of the MDS patterns that we saw in Chapter 4. Agreeableness captures some meaning that would form part of Neuroticism in the five-factor model. Furthermore, Agreeableness could be separated into Agreeableness and Honesty-Humility (using the definitions of the HEXACO model). Yet, this separation only occurred when extracting six factors, suggesting that these two traits are fairly similar. Openness to experience only emerged as a separate factor when extracting five factors and did not correlate strongly with any of the three higher-order factors. Summarizing this rich and complex study, the five and six factors discussed in Chapter 3 can be identified across these languages, but the relevant factors do not strongly emerge in all languages. It appears that people encode and use personality traits somewhat differently when talking about themselves in these mainly European languages.

RELATIONSHIP BETWEEN VALUES AND TRAITS ACROSS CULTURES

So far, we have seen that both values and personality traits show less consistent and coherent structures when examined in Asian, African, and less industrialized samples. In previous chapters, I made the argument that values and personality traits are organized by the same underlying biological processes. This evidence of variability in the organization of values and personality should immediately raise a red flag: is the link between values and personality traits stable across different economic, ecological, and cultural contexts?

To answer this question, I have to go back to the meta-analysis that Diana Boer and I conducted and that I discussed in detail in Chapter 4 (Fischer & Boer, 2015). When we examined the correlations between traits and values separately across the 14 countries from which the samples were collected, we found significant variability. This means that the correlation between values and traits is not uniform, but varies quite substantially across samples and countries. As you might recall from Chapter 4, the strongest correlations between traits and values were observed between Openness traits and the Openness versus Conservatism value axis and between Agreeableness and Self-Enhancement versus Self-Transcendence values. Yet, the estimates for the correlation of values with Openness traits varied between −0.41 (Kenya) to −0.93 (Israel) and for the correlation with Agreeableness between 0.43 (Mexico) and 0.87 (Italy). This is quite a large range. Similar or even greater variability was found for the other trait-value correlations. For example, Extraversion's correlations with Self-Enhancement vs. Self-Transcendence ranged from a near-zero relationship of 0.08 (Hong Kong) to a sizable correlation of 0.56 (Italy). In other words, the relationship between traits and values is highly variable. A closer examination of the pattern again suggests that non-Western samples show lower correlations compared to Western samples.

SUMMARY

The proposed structure of both values and personality that I discussed in Chapters 3 and 4 seems much less stable once we start looking beyond Western countries and examine data from non-Western samples. We covered a lot of ground in the chapter, shining a penetrating, sharp statistical light on the evidence to determine whether there is sufficient evidence to argue that the data are invariant and equivalent. Answering the major questions with which I started this chapter, it could even be questioned whether some of the Big Five traits and values even exist in the minds (and vocabulary) of people in non-Western societies. None of the studies so far has reported full score invariance for either values or personality traits in the Big Five/HEXACO tradition that would allow us to interpret traits or values as straightforward latent variables. Applications of standard instruments in traditional samples from the Amazon and South Asia yielded different factor structures, structures that in turn were replicated within each community. Instruments that were designed from the ground up with non-Western societies typically showed additional or more differentiated traits. These traits were particularly focused on interpersonal and social aspects. This suggests that people in non-Western contexts might describe themselves in different ways than do people in the Western world. This casts some doubts on the universality of the personality and value structure.

Some researchers may claim that I am overstating the evidence. I would like to counter that I am raising questions about existing data that can be addressed with further empirical research. There are three main questions: (1) are there the same number of traits and values across populations, (2) can they be measured with the same instrument, and (3) is the relative organization of traits and values to each other similar or different across human populations? Breaking these down, the first question is whether the very structure of the Big Five and values is similar or different across diverse human samples. This is the question of functional invariance that I raised in the beginning. Can we identify functionally similar sets of behavior that capture the proposed underlying biological systems? A related second question is methodological: if we assume that there is functional invariance but that the behaviors and values that are culturally appropriate vary, we may tailor our instruments to the local samples. This relates to the question of configural invariance. If there is cultural variability in expression, instruments may have different terms and items, but still capture the same basic value and trait dimensions. The lexical approaches suggest that there might be some overlapping dimensions, but also point to additional traits and greater or weaker differentiation of traits in some contexts than in others. If, on the other hand, we assume that the same items or descriptions of behavior or values can be used in our instrument across samples, then we can proceed and ask whether each item or behavioral/value description is related to the presumed underlying factor to the same extent. In this scenario, we can ask more questions about the invariance of items (e.g., do the items and scales meet metric and scalar invariance?). One important further question is whether the relative organization of traits or values is similar or different (e.g., are Extraversion and Openness traits equally strongly related to each other across samples, assuming appropriate measurement in each sample?). This question does not neatly fit into the psychometric questions that I outlined at the beginning. Yet, it is an important question if we want to understand human personality. These three questions can be addressed with existing and new empirical data.

Personality researchers in the Big Five tradition, as well as value researchers in social psychology, typically assume that the Big Five and values in the Schwartz models are basic properties of human nature that exist in most if not all human societies. At the same time in relation to the second question, these researchers are open to the possibility that instruments need to be tailored to specific cultural and social environments. Yet, some of the evidence that I reviewed here suggests that the first question about the universality of the very structure is worth asking. I could also reframe my questions in light of common definitions of traits. Traits – as I showed in Chapter 3 – are often defined as relative stable patterns of behavior, but many personality researchers qualify this claim by stating these behaviors only emerge in response to trait-relevant situations (see for example, DeYoung, 2014; Mischel & Shoda, 1995; Read et al., 2010). We live in a particular type of social and economic environments, and the Big Five might only describe individual variability well in such an industrialized and economically well-off environment. If we define traits in this particular way, we should also be open to question whether individuals in different environments use the same traits to describe themselves.

The third question about the relative organization of the major traits and values is typically not on the research radar screen either of personality or value researchers. Yet this question also merits further attention. It could be easily addressed by reanalyzing existing data sets, in particular the relative organization of Big Five traits and values relative to each other (the proposed organization that I suggested in Figure 6.3).

Where to go from here? I ended the last few chapters very optimistically, with a neat picture emerging that could point to a straightforward comparison of Tina and Sandra (and other human beings from anywhere in the world). This chapter did cast some shadow and doubt on these optimistic approaches. Should I just revise the previous chapters? As I had discussed in Chapter 1, we are dealing with self-reports that rely on a complex set of cognitive processes and heuristics. Maybe I got it all wrong by relying on such self-reports. Maybe the problems that I outlined in this chapter just simply indicate that we are looking at noisy data.

This is one possibility. But on the other hand, if we were to find meaningful processes that could explain these instabilities, variations in reliabilities, and alternative structures, then these problems would be less of an artifact, but rather a systematic source of variation that could be studied and integrated into the model that I started discussing in the previous chapter. This is my challenge for the next chapter – is there indeed some systematic process at work that could help me rescue my nascent model?

8

Understanding Structural Variation: Resources, Threats, and the Power of the Situation

We have to solve a dilemma. On one hand, following many other researchers before me, I have proposed a specific organization of values and personality. These traits and their organization are thought to be evolutionarily stable, and even chimpanzees are supposed to show these traits (Latzman, Freeman, Schapiro, & Hopkins, 2015; Latzman, Sauvigné, & Hopkins, 2016). On the other hand, in the last chapter we saw that the structure is much less universal than commonly proposed. Is this variation just noise and random error, or is there some order in this data mess? How can we reconcile these two positions? In this chapter I address these questions by returning to the discussion of the importance of both the situation and resources. Situational features and availability of resources are crucial elements in the neural network models of personality that bridge neurobiological and social-cognitive models of personality, which I discussed in Chapter 6. We need to bring these features back into the picture of the person to understand what is going on.

RESOURCES AND THREATS: POSTMATERIALISM THEORY

Neural network models of personality (Read et al., 2010; Yang, Read, Denson, Yu, & Pedersen, 2013) explicitly model resources as an important input factor for personality traits. They use a range of person-related resources, including social skills (e.g., face-saving skills, being able to engage in small talk), cognitive skills and competencies (job skills, quick thinking, wit, intelligence, attention span), possessions (money, time), and social features of the person (status). Economic resources have been most extensively studied by researchers.

When we talk about resources in the context of values, we have to mention the seminal theory of postmaterialism and national value change first proposed by Ronald Inglehart (1971), which was later updated and extended to a theory of universal human development (Welzel & Inglehart, 2010; Welzel, Inglehart, & Klingemann, 2003). Its basic argument is that increased access to material (economic) resources within a country will shift value orientations from materialist concerns to postmaterialist orientations, in which concerns for personal growth, the state of the environment, and civil liberties become more important. Translated into the terms used by Schwartz, values shift from conservative and self-enhancing values toward open and self-transcendent values over time. The theory proposes a sequential process that shifts these value preferences with the increasing availability of economic resources, via transformations in life strategies. Different life strategies are deemed effective in different living conditions (which vary both between individuals and societies). In line with broad evolutionary arguments of adaptiveness, Welzel and Inglehart (2010) focused on the capabilities to satisfy the needs of survival and reproduction and to use opportunities in the environment to thrive. They argued that individuals and societies can be ranked to the extent that they promote individual agency. In conditions of scarcity where life strategies are focused on pure survival, there is limited agency to achieve personal growth. Once environmental conditions become more favorable and existential threats to survival recede as a focal goal, personal agency focused on personal development can develop.

The crucial element that sets this process into motion is the availability of resources to individuals made possible by socioeconomic development (Inglehart & Baker, 2000). Socioeconomic development is a broad term that includes processes as distinct as increasing wealth, occupational differentiation (e.g., more diverse jobs, especially in the service industry, which go beyond jobs in agricultural production and industry), increasing education levels, and urbanization. How do these processes influence agency? Wealth is associated with increases in income, which allow the average individual to meet survival needs better and more efficiently. Occupational differentiation and increasing education levels provide more possibilities for skill and information acquisition and utilization. Urbanization and increased social mobility allow individuals to emancipate themselves from closed traditional in-group cleavages, which then decreases vertical power hierarchies and dependencies, thereby opening opportunities for horizontal exchanges among more equal citizens. All these diverse processes in conjunction increase human choices via access to enlarged individual resources.

These processes capture some of the resources modeled by Read and colleagues (2010) in their simulation of personality dynamics, which I discussed in Chapter 6.

The initial postmaterialism argument and empirical evidence showed that increasing socioeconomic development is associated with a shift in mean value priorities. This process can therefore potentially explain the variability in means across nations that we saw in Chapter 5 (I will review and extent this work in the next chapter). However, Inglehart, Welzel, and other researchers in this tradition did not consider the possibilities that economic resources may influence the very structure of values (or traits).

Fons Van de Vijver and Ype Poortinga (2002) were the first to draw to attention to this potential implication. They analyzed the postmaterialism value index with data from 39 nations ($N = 47,871$). When examining the factor structure of this widely used measure, they found a single factor, in line with previous research by Inglehart. However, the factor congruence varied substantively across these 39 nations in their analyses, ranging from 0.89 to 0.99 (with Poland being an extreme outlier, probably due to social upheaval during the period of data collection). When the authors correlated both these factor congruence coefficients and internal consistency estimates with national wealth, a significant and positive correlation emerged: in societies with greater income overall, individuals differentiated more explicitly and clearly between materialistic and postmaterialistic values.

This may sound surprising at first, but it is actually quite compatible with the overall postmaterialism logic. The central argument is that agency develops with increasing individual resources that become available when socioeconomic development increases. People have more personal choices, because having access to more cognitive and material resources implies that restrictions on personal choice are increasingly removed and people can decide what they want or do not want. This opens up possibilities to choose different life trajectories.

In conditions of scarcity, there are few options beyond pursuing any means that will ensure the survival of oneself and one's closest kin. Keeping good social relationships is instrumental for meeting basic necessities. Maximizing material benefits is important, and displaying status via accumulated wealth is adaptive to ward off competitors and increase the potential to attract high-quality alliance partners. Hence, the choice of different values (as life goals) is restricted in a real and existential sense. Individuals choose values in an instrumental fashion, so that each choice they make is determined by the extent that pursuing a particular value will allow them to meet needs of survival and reproduction. This should result in a reduced cognitive differentiation of values: everything is valuable to the degree that it helps you to survive.

In contrast, if the immediate threat of not meeting resource needs is reduced and therefore concern about immediate physical survival is lessened, individuals can use their energy and time to consider other options in life. Similarly, display of wealth is less important as a status signal because of weakened hierarchies, and the desirable qualities of a potentially attractive mating partner or ally include more than material resources. Options become available to pursue other goals and life strategies. Taken together, this process leads to a differentiation and elaboration of values.

ECONOMIC RESOURCES AND THE SCHWARTZ VALUE THEORY

Fontaine and colleagues (2008) were the first to systematically examine the stability of the Schwartz value model. In the previous chapter I reviewed the structural part of their study, in which they found that the prototypical structure of values showed considerable variation across samples. They also attempted to explain the deviations from the originally proposed theoretical model. A first and important source of variation that they considered was simply random fluctuations due to sample size. It is well known that, in smaller samples, estimates randomly oscillate more because they are based on less information, and therefore, variability of estimates is higher. Fontaine and colleagues concluded that about 14% of the variability across samples could be attributed to these sampling fluctuations in their study. This is quite a substantive amount of random error. My concern in the previous chapter that we might be looking at noisy data is not unfounded.

A second source of variability, but this time of systematic instead of random variability, was whether the respondents were teachers or students. This accounted for about 5% of the variability. Student samples typically showed a clearer structure in line with the theory compared to teacher samples. The major distinction was that the younger student samples differentiated more between personal and social values.

Finally and most importantly for my considerations here, socioeconomic development had a significant and large effect. Fontaine and colleagues used a composite index that included indicators related to economic conditions, education, occupational diversification, health, access to mass communication, level of democracy, and population dynamics, hence capturing a global index of postmaterialism. About 39% of the variability was explained by this index (and another 7% in interaction with respondent type: effects of socioeconomic development were more pronounced in teacher samples).

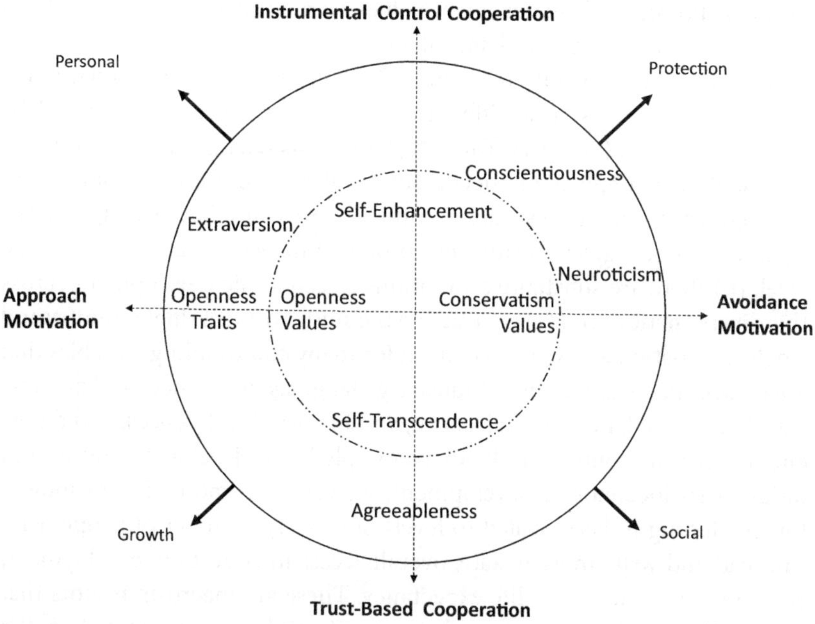

FIGURE 8.1 The conceptual trait-value model with the rotated dimensions by Fontaine et al. (2008) included.

This effect was therefore much more substantive than the variability due to random fluctuations. The problems with structural instabilities that I identified in the previous chapter might be amenable to scientific study after all.

To understand the effect of socioeconomic development, Fontaine and colleagues investigated the effect on the two major dimensions separately. They rotated the main axes of the value circle slightly differently from the two major axes that I discussed in Chapter 3. One dimension differentiated between person-focused values (Achievement and Hedonism) and social-focused values (Tradition, Conformity, and Benevolence). Their second dimension differentiated between protection values (Power and Security) and growth-focused values (Universalism and Self-Direction). Figure 8.1 shows these dimensions rotated at a 45-degree angle in the conceptual model from Chapter 6. The authors found that socioeconomic development particularly affected the differentiation of values along their second dimension. Growth and protection values became more differentiated (correlated more strongly and negatively) in more socioeconomically developed nations. Analyzing the data at the item level, a large number of

Security, Power, and Achievement values were located farther away from Self-Direction and Universalism values.

Together with several colleagues (Fischer, Milfont, & Gouveia, 2011), I tested this idea using a different value theory developed in Brazil by Valdiney Gouveia (Gouveia et al., 2014). We collected data from a nationally representative sample of physicians across all 27 states in Brazil. Brazil is an excellent case study to test these economic processes, because (a) its large number of states gives us sufficient power to detect any effect at the state level, (b) there are substantive variations in economic development across this diverse nation, and (c) conducting the study within a single country and single occupation allows us to control for many confounding variables that are present in cross-national studies (e.g., language differences and possible translation problems; cultural, social, and historical differences; and differences between groups of individuals sampled). To develop a good overall index of socioeconomic development, we used a combination of indicators, including indices related to levels of poverty, number of people who can read and write in each state, overall access to education, employment, economic inequality, and life expectancy. These are macro-indicators that capture important economic and structural conditions within a state. We were also interested whether living conditions of individuals might have a detectable influence. After all, socioeconomic development often means a change in living conditions – as people gain more resources (higher income, better education, and good jobs), living conditions improve, allowing individuals to live separately from their extended family and to pursue their own interests independent of their immediate family. Such an independent lifestyle is also associated with having fewer children (because other goals in life might become important). To capture all these dynamics, we used a combined index of the percentage of married couples living with both their children and parents in the same household, the number of births per capita, and the ratio of children to retirees within each of the Brazilian states.

Replicating findings by Fontaine et al. (2008), we again found a substantive effect of sampling, with smaller samples showing more variability than larger samples. Second, also replicating Fontaine et al.'s findings about the difference between students versus teachers, samples that had a lower mean age showed a clearer structure compared to samples with an older average age. This shift was particularly strong for the distinction between social and personal values. Importantly, socioeconomic conditions had a sizable effect on the value dimensions, with living conditions having an even stronger effect. The socioeconomic effect was mediated by living conditions, because once living conditions were included in the model, economic variables were not important anymore. Doctors living in states with crowded living conditions were less likely to differentiate their values than those living in less crowded states. Interestingly, this effect was stronger when considering where individuals were born and grew up, suggesting socialization effects early on in life. These studies were conducted with value ratings in the traditional form, which asked for the most important values in life.

A number of studies have demonstrated these effects with PVQ ratings, which use a format that is much closer to trait ratings. These studies were conducted in representative samples across Europe (Bilsky, Janik, & Schwartz, 2010; Strack & Dobewall, 2012). This means that these findings are good reflections of what is happening within European societies overall. I focus on the most recent analysis by Strack and Dobewall, because their research goal was more closely aligned with the purpose of the current chapter and they used sophisticated methods for analyzing the effect of economic development. The authors used representative data from 33 countries collected in multiple waves of the European Social Survey (resulting in 98 samples). Using factor analysis, they first replicated the two-dimensional structure of the PVQ (with the dimensions being in line with the classic model differentiating Openness to change versus Conservatism and Self-Enhancement versus Self-Transcendence described in Chapter 3). Second, replicating the problems that I discussed in the previous chapter, there was substantive variability in the explained variance of each dimension in the data. In other words, the two dimensions did not work equally well everywhere.

To account for this variability, Strack and Dobewall (2012) constructed a socioeconomic development index consisting of income per capita, percentage of people in vulnerable employment, percentage of females elected to parliament, subscriptions to broadband internet, and how many people were employed in research and development positions (all values adjusted to account for population size). They then rotated the two axes to align with the alternative dimensions described by Fontaine et al. (2008). The growth versus protection dimension (Universalism and Self-Direction vs. Security and Power) was strongly related to socioeconomic development. In nations with higher development, the growth versus protection factor became stronger and explained more variance. In less economically developed European countries (e.g., Hungary, Romania, Slovakia, Turkey, and Ukraine), the factor in some cases even disappeared completely, leading to a collapse of this dimension and a unidimensional structure of values. Figure 8.2 shows this process schematically.

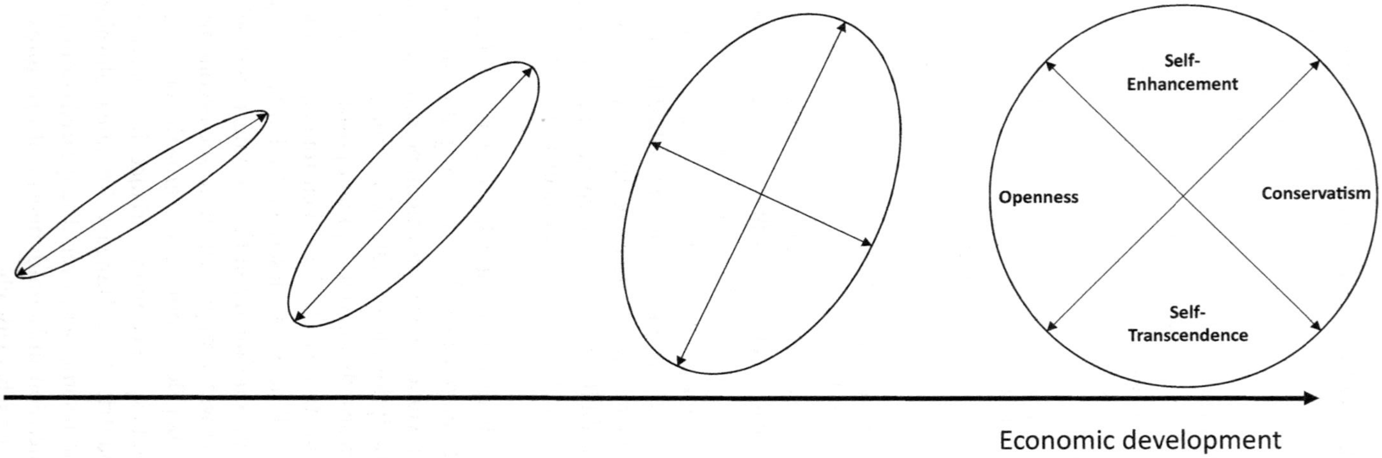

FIGURE 8.2 A schematic depiction of the gradual elaboration of the value structure with increasing societal wealth.

INTEGRATING RESOURCES WITH THE BEHAVIORAL MOTIVATION-COOPERATION FRAMEWORK

In Chapter 6, I proposed a model differentiating approach-avoidance motivation behavior and cooperation strategies to account for the joint organization of values and traits. The growth versus protection dimension is rotated at a 45-degree angle from the two major dimensions (see Figure 8.1).

I just reviewed evidence showing that the problems with the data outlined in the previous chapter are systematic variations caused by differing economic development trajectories across societies. The underlying universal human development model described by Inglehart and Welzel is theoretically important and relevant for the structure of values. In conditions of scarcity, the structure is compressed into an ellipse, separating what Fontaine and colleagues have called social versus individually oriented values. However, I think this labeling is misleading because personally oriented values (Hedonism and Achievement) are still distinctively social. Achievement is about showing competence and expertise according to socially sanctioned standards, whereas Hedonism refers to enjoying a good time. Given the hypersocial nature of humans, having a good time typically involves spending time with others that are significant in our lives. Hence, both value types are very social in focus. Therefore I believe it is better to think of this compression into an eclipse in terms of an interplay between approach versus avoidance motivation and the strategies of cooperation that are adaptive in a specific environment.

If people have sufficient resources to pursue their own interests instead of being preoccupied with survival, they can approach novel situations and explore opportunities in the environment. Therefore, this lack of survival threat should be associated with a decreased activation of the behavioral inhibition system (BIS) and an increase in the activation of the behavioral approach system (BAS). Simultaneously, individuals also feel more secure in their dealings with others and may be less worried about being taken advantage of. Therefore, individuals do not need to very carefully monitor their resources and have less of a need to use social relations to advance their own position. Hence, trust-based cooperation becomes a possibility. In contrast, in situations of scarcity, individuals are primarily concerned with survival. First of all, such situations may increase the behavioral inhibition of individuals. They may switch goals to pursue paths that are well trodden and that are therefore more likely to bring results, instead of approaching new situations and experimenting. Using models from evolutionary psychology, prestige-based transmission biases – copying the behavior of individuals who appear rich and powerful in one's environment (just think of the large numbers of people who copy the behavior of stars and celebrities in today's world) – should increase (Mesoudi, 2009; Mesoudi et al., 2006), and associated conservative values become more prevalent. Living with scarcity, individuals should be more focused on using social cooperation to advance their own standing and survival, with cooperation being more instrumentally driven. Because this concern of survival is more important, the relative contrast of these values along the growth versus survival dimension becomes weakened. Yet it is not an independent dimension, but the interplay of these major biological systems (the interplay between BIS and BAS versus the mode of cooperation system) that decreases the motivation of individuals to separate these particular values.

In situations of scarcity, the relative distinction of Achievement (and Hedonism) versus Benevolence, Tradition, and Conformity is still valid because they offer distinct, but viable, paths toward securing survival. The distinction is driven by the properties of Achievement along this axis. Achievement is likely to be activated by both the BIS and BAS, allowing the cautious approach of opportunities if they appear socially safe. In contrast, the conglomerate of Benevolence, Tradition, and Conformity values is most likely more strongly associated with the BIS, involving an inhibition of any impulses that upset others. Benevolence involves trust-based cooperation, but with individuals who are within one's immediate social environment. Therefore, the overall net effect is that relying on traditional and trusted social cleavages provides a path for securing sufficient resources for survival. Therefore, these contrasting value clusters offer two possible options for survival that are differentially activated by the major two underlying systems. This feature distinguishes these values from the other rotated dimension (growth vs. protection). The combination of behavioral approach motivation with trust-based cooperation (relating to Universalism and Self-Direction as growth values) is not an optimal strategy for survival in conditions of relative economic deprivation. Hence these values become less important to the overall structural organization, they become mingled with other values, and this in turn leads to the emergence of a single value dimension only.

In summary, the availability of resources within an environment interacts with the two major dimensions that underlie the organization of values. In contexts of economic scarcity, the life strategies (borrowing from postmaterialism theory) change, and the value space becomes cognitively simpler. Making more fine-grained distinctions between values is not very adaptive, and therefore, only the most fundamental value distinctions between an emphasis on traditional social groups and cleavages versus socially approved exploration of the environment are salient in people's minds. It is the complex interplay between behavioral approach and avoidance motivations with modes of cooperation that can explain this cognitive narrowing of life goals. When resources become more plentiful and available, individuals are less pressed to ensure their survival, and therefore, they can make more fine-grained differentiations as conditioned by their underlying genetic predispositions. Genes do not trump the environment: they interact with the environment, creating a conditional if-then relationship.

RESOURCES AND PERSONALITY TRAIT STABILITY

So far I have focused only on values, primarily because values have been most researched in this regard. What can we say about traits? In the previous chapter I discussed both congruence coefficients from a large study using students to report on the personality traits of others whom they knew well (McCrae et al., 2005) and reliabilities from a study that used the HEXACO in student populations (Thalmayer & Saucier, 2014). One straightforward option is to examine if there is a relationship between these structural parameters reported by the original authors and economic development. Let's have a look.

I took averaged national wealth per capita from the 1970s to 2006 from a previous study as a general indicator of societal wealth (Fischer & Boer, 2011). This index captures the overall economic conditions for the last 40 years, which should be sufficient for evaluating the impact of economic conditions on individuals who provided the personality ratings in these various studies. The reliabilities for the HEXACO self-ratings across 26 nations showed significant correlations with this wealth index for four of the six HEXACO dimensions. The strongest correlations were 0.62 for Emotional Stability and 0.61 for Extraversion. Reliabilities for Agreeableness ($r = 0.52$) and Openness ($r = 0.50$) also strongly and significantly correlated with national wealth. The reliability for Humility reliabilities did not significantly correlate with wealth (0.26). The weakest correlation was with Conscientiousness (0.10). Overall, the higher the national wealth, the stronger were the reliabilities. This implies that personality traits become more coherent and meaningful for participants in environments that are economically developed.

A more direct test of the similarity of factor structures is an analysis of the congruence coefficients reported by McCrae and colleagues (2005). All correlations between the congruence coefficients and the index of societal wealth were significant ($p < 0.05$). The strongest correlation was with Openness (0.53), followed by Agreeableness (0.51), Neuroticism (0.45), and Extraversion (0.42). The weakest, but still significant, correlation was with Conscientiousness (0.36). This clearly demonstrates that traits become more similar to the prototypical structure found in the United States in national contexts that are more economically developed. In other words, the same economic variables that affect the structural organization of values also affect how stable and coherent personality traits are, when measured with two different personality inventories across a large number of samples.

The Conscientiousness trait was least affected by economic development in both samples, which is in line with the results of the lexical studies reported by De Raad and colleagues (De Raad et al., 2010, 2014). They had suggested that Conscientiousness is a universally important personality trait and is likely to emerge independent of local social, cultural and economic conditions.

These results, unfortunately, do not tell us much about the relative organization of the traits to each other. There is much scope for future studies to explore this relative organization (see Figures 4.7 and 4.8) and how this organization might change depending on the economic conditions of the population. In fact, if these separate correlations were consistently reported in published studies, a curious reader could reconstruct and test these ideas directly. Not having access to this data, one potential way to assess the relative ordering of traits is to reexamine the correlations with values and the extent to which value-trait relations are affected by environmental conditions.

VALUE-TRAIT CORRELATIONS AND CONTEXTUAL THREAT

To gain insight into the relative organization of traits (in relation to values), I return to the meta-analysis of value-trait correlations that I discussed extensively in Chapter 4. As you might recall from that chapter, values and traits on average are consistently correlated. In Chapter 6, we also discovered that this relationship varied across samples. Let us examine whether this variability is systematic and can provide further insights into the underlying processes.

In our original article on value-personality correlations, we tried to account for these differences (Fischer & Boer, 2015). We thought it might be more interesting to focus on threats instead of resources, because it is the lack of resources that is likely to produce these psychological effects. Therefore, we created indicators of threat, with threat being defined as a lack of resources and issues that threaten the survival and well-being of individuals. A first threat that we included was economic – lack of money as an economic resource. This is very much in line with what I have been discussing so far in this chapter. In our study, we used gross national income per capita as an indicator of average wealth available to individuals. We also included two other forms of threat. First, we computed a broad indicator of ecological threat, which included the amount of communicable diseases prevalent, average amount of years lost due to diseases, the amount of calories that people consume per day (which we inversed to capture hunger), polluted air, and death rates due to environmental factors. Another type of threat was related to institutions and the general social environment. Here we included crime rates, as well as restrictions to press freedom, freedom of travel, labor movement, and political participation. These three types of threat were highly correlated with each other, suggesting that they capture an overall environmental context in which people face multiple threats and have limited resources to enable a comfortable living.

Overall threat moderated the relationship in 7 of the 10 tests we ran. The pattern is that increased threat is associated with weakened value-trait correlations. Values as motivational goals and traits as dispositional behavioral tendencies only correlate if individuals have the necessary resources and do not feel under threat. Threats on average explained at least 10% or more of the variability in the strengths of correlations.

Assuming that values and traits are linked to the same neurobiological system as I have proposed (remember Figure 6.3), the availability of resources to meet contextual threats is systematically affecting how closely values and traits correspond. As we have seen, lack of resources also affects how well values and traits are organized in the minds of people. Therefore, the extent to which these two psychological constructs are clearly organized and aligned with each other is affected by the available resources. Thus, the balance of resources in relation to the contextual threats may affect personality as an overall systematic organization of psychological constructs within an individual at a more profound and existential level than just shifting the priority of traits or values. Resources and threat appear to affect the very organization of how a person thinks, plans, and behaves. With this thought in mind, let's have another look at the role of the situation as discussed by Read and colleagues (2010).

REEXAMINING NEURAL NETWORK MODELS OF PERSONALITY: THE IMPORTANCE OF SITUATIONS AND BEHAVIORAL OPTIONS

In Chapter 6, I introduced neural network models of personality as a bridge between the biologically and the social-cognitively focused models of personality. Building on these two different research strands, Read and colleagues (2010) conducted a simulation study that manipulated situational features in a number of ways (I give some examples in parentheses). They varied whether a situation was work, home, or leisure related; whether and how many other people were present; the status of these individuals (kids, relatives, friends, subordinates, boss, romantic partner or potential date, as well as age differences between individuals); the nature of the situation (busy at work, conflict situation, during break); location (restaurant, bar, at desk); and other potentially relevant features (alcohol, dancing, TV present). The important insight gained from these manipulations was that situational features strongly influenced behavior. As you may recall, Read and colleagues started off with two basic biological predispositions: differences in situational features helped differentiate the behavior approach system and the behavior avoidance system into the emerging traits of Neuroticism, Extraversion, and Conscientiousness, depending on the manipulated goal settings of the underlying BIS and BAS. Hence, the situational features interacted with biological predispositions and led to the emergence of traits that we recognize as being part of the Big Five.

Zooming out and looking at their manipulated situation from a more global perspective, these situational features are very much geared toward an environment that is industrialized and Westernized. It involves office work and dealing with work colleagues, immediate family, and a circle of friends that we choose; some recreational activities involve alcohol and dancing. This choice of situations is not surprising given where the authors were located and what they aimed to model (personality traits in Western samples). At the same time, these features do not capture the situational realities of a large proportion of humans. For starters, North American and Western European individuals predominantly live in nuclear families and typically have a small set of people (their partner and maybe one or two children) with whom they need to interact on a daily basis at the family level. Outside the immediate family, Western individuals may chose to socialize with people around common interests: sports, arts, other hobbies, or political affiliations. This means they have options of whom they want to interact with. These interactions require a specific set of interpersonal and social skills in attracting and maintaining social relationships.

In contrast, in many parts of the world, people have extended families – relations need to be managed with in-laws, grandparents, uncles, aunties, nieces, nephews, cousins, and so on. Compared to many Western contexts, it is not possible to leave these relationships easily: people are embedded in these relations with few safe options of leaving this network. Living conditions are also often more crammed, and people have no choice but to interact with many others on a daily basis – recall the study in Brazil where we found that living conditions had an influence on how well differentiated values are for highly educated doctors. These effects that we found for medical doctors are likely to be more pronounced among those who have less resources and lower levels of education. Working conditions also vary dramatically – even in the Western world. Office work might require different skills than working on a farm or in a factory or sweatshop. How people socialize after work and in their social networks also varies dramatically. For example, situations involving alcohol are forbidden to large proportions of humanity for religious reasons. What are the implications of these different living and social conditions for personality?

Read and colleagues (2010), among other personality psychologists (Corr et al., 2013; DeYoung, 2014; Mischel & Shoda, 1995), have argued and empirically demonstrated that traits become manifested if they are triggered in a specific situation. In fact, traits are often defined in relation to situations that allow these traits to emerge. Extending this argument, we can easily imagine that the different situations that structure the lives of people around the world may very well lead to the different expression of dispositional tendencies. Daniel Nettle and colleagues (Nettle, Buss, & Hawley, 2011) discussed two different correlational processes that may lead to the emergence of a particular trait structure. First, different behavioral tendencies might converge into a coherent behavioral package if the environment simultaneously favors two different, but independent traits. The authors called this the **correlated environment hypothesis**. They cited an example of a freshwater snail that either grows a small, rotund shell slowly (and gains the ability to leave the water) or grows a long and narrow shell quickly (but is unable to leave the water). This divergence in growth rate, shape, and behavior was driven by the presence of either predatory fish or crayfish in the environment. Snails were more likely to survive in the presence of predatory fish if they had round and hard shells that the fish could not easily crush. Slow growth was important to build strong shell walls. If predatory fish were present at a particular moment, the snail could escape by briefly leaving the water and waiting until the predators disappeared. In contrast, in areas where crayfish are common, the best strategy was to grow long and thin shells. Crayfish cannot crush the shells, but they can enter the aperture of the small round shell to attack the snails. Growing slim and long shells made evolutionary sense because it made it harder for crayfish to enter. Growing long shells in a short period of time also allowed greater options for the snail to retreat if a crayfish was trying to enter the shell. Both strategies as a package of unrelated behavioral features make evolutionary sense.

Nettle and colleagues (2011) discussed a second strategy called **synergistic behavioral correlation**. I alluded to this mechanism in earlier chapters when discussing the convergence of different trait aspects into a broader trait. The idea is that several behavioral strategies in combination, although initially independent, make an evolutionarily more adaptive package. For example, boldness in foraging and aggression might work together. If an animal is bolder in foraging, it may capture larger and a greater number of pieces of food. However, having these food packages might attract other animals, which then requires more aggressive behavior to defend the food against competitors. Hence, independent behavioral traits may converge to form behavioral packages in specific environments that provide an advantage to the individual exhibiting both traits (Sih, Bell, & Johnson, 2004).

Gurven and colleagues' (2013) work that I discussed in the previous chapter showcases a beautiful example of how individual traits may act synergistically in specific environments, leading to the emergence of a different covariation of traits (meaning a different personality structure). In small-scale societies relying on hunter-gatherer and foraging subsistence modes, people live in small bands of extended kin and have limited choice about whom to socialize with or who to attract as a potential partner. They also have more limited options for demonstrating social success and status. Success according to Gurven and colleagues is primarily defined in terms of food production and being able to provide for one's family. As discussed in the previous chapter, they found two major personality trait factors: industriousness and prosociality. These two factors may define two different strategies for demonstrating success in this particular ecological and cultural context. Industriousness appears to be directly linked not only to specific behavioral tendencies to gather more resources and thereby produce more food but also to sharing these resources more generously and widely with others. This is likely to increase social status and attractiveness to potential reproductive partners. Different unrelated strategies that are located on different personality trait factors in Western samples emerge as a coherent behavioral package that signals success among the Tsimane. As a consequence, Tsimane in this traditional community have a different mental representation of what constitutes a successful person.

At the same time, the Tsimane social structure is highly egalitarian. As a result, there are constraints on boasting and broadcasting success (a trait often associated with Extraversion). Because of these various pressures coming from a highly social but more egalitarian environment, where groups are more permanent and leaving existing groups is not an option for individuals, socially oriented traits might be differently structured. An extension and proliferation[1] of socially oriented personality traits have been reported in both Asian (Cheung et al., 1996) and African contexts (Nel et al., 2012). These studies found that, in addition to Agreeableness and Extraversion, other traits with strong social connotations emerged. Such traits often focus on maintaining social harmony, saving face, a greater attention and orientation toward one's family, and a concern with social relationships in general. Similar processes of more finely differentiated and more extensive social value domains have been reported from the Philippines (Fischer, 2013b). Among the Tsimane studied by Gurven and colleagues, Agreeableness and Extraversion items merged to form a particular cluster of prosociality. This extension and elaboration of social relationship-oriented traits appear to be adaptive in environments that differ dramatically from the ones with which we are most familiar.

AN INTEGRATION

In the beginning of this chapter, I discussed postmaterialism as a theory that has been used to explain how values might shift with increasing economic development. Yet, the same mechanisms might be operating at a more fundamental existential level, affecting how people see themselves and others. As pointed out by Gurven et al. (2013, p. 366),

> It is an intriguing possibility that pan-human reaction norms shape not only intersocietal differences in average personality scores but also the structure of personality covariation itself, due to sustained socioecological differences across human populations. This hypothesis cannot be rejected in light of recent cross-cultural studies finding universal evidence of the Big Five, given the WEIRD-ness of most of the study populations. Indeed, any model of personality that specifies a fixed set of biologically based trait dimensions would be inconsistent with the results we report here. A comprehensive theory of personality would need to explain how particular conditions might lead to different combinations of calibrated and coordinated items, which then generate multidimensional personality structure, in varied socioecological settings and circumstances.

[1] As I discussed in Chapter 1, there is a general distinction between theoretical splitters and lumpers. The cultural studies mentioned here focused on enlarging the trait and value space and differentiating more traits and values. A critique of differentiating additional socially oriented values and traits is that these new clusters of traits and values can be subsumed under already existing labels (e.g., Agreeableness). My view is that basic predispositions might be differentially elaborated and expressed in different social, economic, and ecological settings. The interesting question for me is under what conditions do people start making more fine-grained distinctions.

My aim is aligned with these directions. In Chapter 6, I proposed that two major systems organize human personality: one major system separating approach versus avoidance motivation and a second one that organizes modes of cooperation strategies. These two systems interact with the external environment. My proposal is that the threats and opportunities that are present in the environment form the background within which these two major systems provide an individual with behavioral options. The major point highlighted by Gurven and colleagues is that different combinations of strategies are adaptive in different environments (a point echoed by Singh et al., 2013). The environment has a major impact on what values and behavioral dispositions are salient and – importantly – how they are organized in relation to the demands and pressures present in the environment.

I will discuss developmental pathways later, but I want to foreshadow an important point here. Most likely, specific strategies are activated very early on during the development of a person, during early infancy and childhood. These trajectories can still be altered by triggers later in life, but a basic parameter space of the system of a person might be set during infancy. Classic research has demonstrated that maternal care provided by rat mothers activates or suppresses specific behavioral responses (Champagne, 2008). The underlying motivation and cooperation systems might be calibrated early during the neonatal and early childhood period, setting the individual up with the appropriate response strategies to deal with the environment.

In Chapter 5, we saw that there is systematic variability in trait and value levels in populations globally. Some values and traits are rated as higher or more important in some contexts, and others are rated lower or as less important. There is an up- and down-regulation of behavioral responses, resulting in mean-level changes. This is well known, and in the next chapter, I will examine these processes in more detail. However, this should not distract us from the insight that a more fundamental process is shaping the very structure of personality. Resources, threats, and the demands and opportunities present in the environment shape not only the relative importance of values and traits but also their very organization. Economic resources appear to play a major role and are linked to a number of additional processes, many of which might drive socially focused behavioral and motivational features.

"Personality as a City" Metaphor

To visualize this simultaneous organization by level and structure, imagine a small city in a fantasy world. The city has an organic structure and expands and shrinks with the state of well-being of its inhabitants and the environment. There are defensive structures: a prison and a city wall. There are community halls for people to meet, banks and financial institutions, museums, art galleries, and places of worship.

External threats and the availability of food and water change over time, creating various challenges for the city's inhabitants. For example, a drought caused a poor harvest, and water is running low. The inhabitants of a neighboring city-state are also threatening to invade the city to capture much-needed resources. Meanwhile, within the city there is the potential of unrest due to the prevailing scarcity. To cope with all these problems, the city council is raising the height of the walls encompassing the whole city and expanding the size of the prison; it now towers over the city like the Bastille to remind citizens of where they might end up if they were to disobey the law. At the same time, art exhibitions, museums, and schools either are closing down or are reduced in size. A fantasy bird flying above would see a city with small and narrow streets, meager little houses nestled together, and everything surrounded by that tall, big, fat city wall. The prison tower looms large over the small houses, threatening in its menacing stare over the city. The warehouse and water tanks storing the precious resources are heavily fortified and rise high above the roofs of the houses in the city. A giant church-like structure in the middle of town regularly demands citizens' attention and calls them to order and pay respect. Little oases of the arts and science are found only in rundown small buildings, barely noticeably and dispersed across the city. Defensive structures and places housing resources are overpowering and tall, with all important parts of the city squeezed together in small and cramped neighborhoods. This is a structure under threat and faced by a lack of resources.

Yet, the city's fortunes improve. There are plentiful harvests in the years to come, the neighboring city-state again engages in friendly relations, and plenty of food is available for everyone. People level the city walls, and buildings start spreading into the open fields. An artist colony appears in vast green spaces far from the old oppressive city center. Tall museums and art galleries are erected. In a new neighborhood in the south, a big community center is built, with colorful flags flying from a tall tower, indicating the importance that good neighborly relations now play in the life of the city. Small shopping centers open in the far north, but they are less oppressive in comparison to the dark towers that once dominated the city center. Now our fantasy bird is seeing a dispersed and bright city. Different neighborhoods specializing in different activities are clearly visible and distinct. The places that are most important now and therefore have the biggest, tallest, and brightest buildings are facilities for the arts, centers of learning, and places of communal get-togethers.

I use this metaphor of a city to illustrate that both the overall structure and relative importance of individual behavioral traits and values within the "city parameters" of a person are changeable across environmental conditions. There is plasticity in the overall system, and the environment strongly influences how the individual parts are operating together and which part is given greater weight in the organization of a person. The arts and culture district is equivalent to Openness traits and values. The large community center symbolizes Agreeableness and Self-Transcendence values. The oppressive defensive structures, prisons, and places of authority are Conservatism values (and to some extent Neuroticism and Conscientiousness). The towers of commerce reflect Self-Enhancement values, as well as Extraversion and Conscientiousness.

This metaphor provides a visualization of these processes. If we assume this metaphor as an empirical possibility, it has wide-reaching implications for how we understand personality processes. Here is just one prominent example that has resonated in various forms in the media over the last few years. How will the Syrian refugee child who is growing up in a camp in Lebanon or Turkey understand herself and others around her? What are the implications for evaluating this person for a job (if she was ever lucky enough to escape the camp and seek formal employment)? How can we assess her well-being if she needs psychological treatment? What educational incentives based on what models of personality would help her grow? These are broad questions that are not often addressed, given the widely held assumption that personality systems like the Big Five or the value circle are near universal. This issue will concern us in the final chapter again.

Research Questions and Implications

Supposing that this city metaphor applies to humans, what are the implications for theory and research? In this chapter I treated resources, threats, and situations as big unitary concepts, with a lot of unpackaged meanings. Ronald Inglehart and Chris Welzel's discussion of economic development encompasses many different processes that might unfold sequentially and at different scales for individuals and communities. To take just one simple example, changes in economic conditions are associated with differences in access to education. With more income, more individuals can afford education. Simultaneously, more economically active regions need more individuals with higher education to keep the economy going. Are the effects that I discussed throughout the chapter a result of having access to more goods, or do the changes in education and the different schooling and vocational training lead to different mental abilities that are necessary to thrive in a more economically active environment? National wealth and any of the other variables (levels of schooling, democracy, unemployment, social inclusion, health statistics, population growth, and so on) that were included in the studies that I reviewed in this chapter need unpackaging. What are the mechanisms in this package of economic development that are most important for understanding these processes at the psychological level of the individual? There is a wide scope for future research to dig into these dynamics and their relevance for personality. We need to differentiate the relative importance of different variables encompassed under the umbrella of overall resources and threats. A related avenue for further research is a closer attention to temporal sequences, to how these processes unfold over time. What variables are most important at what time in the life of a person and as communities develop economically?

The situational features may also benefit from further unpackaging. What features of the situation are most important for personality and value development? Socially oriented traits and values might become more differentiated or follow a different organization in environments where people are embedded in cohesive and enduring social groups. Social skills vary depending on whether you can choose your friends or you mainly have to stick with those people who are part of your social network through kinship and social relations. What elements of the social system are most important – the more extended contact with different caretakers in large social networks versus the more intimate contact with your biological parents in most Western societies? What is the influence of smaller families on personality and values? How do school and work environments (large classrooms/industrial complexes vs. small study groups/autonomous teams; competitive learning and testing environments) and the networks that people develop at school and work relate to personality traits and values?

A second major question and implication concerns the status of latent variables and measurement invariance. Much of the personality literature implicitly or explicitly assumes that there are latent variables that are expressed in ratings of items in personality inventories. This assumption is also in line with the widespread use of factor analysis, which assumes that there is a latent unmeasured variable that causally influences observed responses. At the same time, in the previous chapter we saw that to date it is very difficult to find metric or scalar invariance in cross-cultural data that would unambiguously support the statistical existence of a *single* latent variable that is sufficient to explain the covariation among questionnaire items.

In line with neural network models of personality, I assume that there are at least two (or three – if we keep behavioral approach and behavioral avoidance motivation separate) independent biological systems that interact with each other and environmental conditions, which in turn then lead to the emergence of varying sets of traits and values that are adaptive in specific environments. This argument implies that there are no latent factors at a psychological level, because independent biological systems in combination with the environment shape how individual behave (and how they subjectively construct their values and personality traits deals with ratings and/or lexical descriptions). Therefore, it will be impossible to find scalar invariance, since there is always more than one single underlying factor involved in any one of the Big Five or any of the 10 values.

The proposal that underlying biological systems interact with features of the environment may require different factor specifications. It might be possible to specify (a) more complex statistical models or (b) psychological latent variables that satisfy the assumptions of a latent variable model. Bi-factor models (Holzinger & Swineford, 1937) have become more popular in the personality literature. In these models, a researcher proposes an independent factor that is set to influence all observed scores, but that is unrelated to the (for example, Big Five) content factors. The interpretation of such a general factor is controversial, because it might capture method artifacts or it might indicate a general evaluative or affective factor (Bonifay, Lane, & Reise, 2016; Rodriguez, Reise, & Haviland, 2016). Instead of having a single factor, we could theoretically specify separate behavioral approach versus behavioral avoidance motivations and type of cooperation factors that then jointly affect items (maybe in addition to the Five Factors). Such a model is nevertheless likely to run into statistical problems (because we need to estimate many more parameters, and doing so creates problems for model identification and stability). Such a model would also not capture the influence of the environment. The second option is to specify latent variables of a psychological kind. In this case, the researcher would need to specify how these latent variables are operating (are they related or instantiated via specific mental structures or brain networks, how do they relate to biological and environmental variables, etc.). Denny Borsboom (2006, 2008; Borsboom et al., 2003) discussed the theoretical questions that researchers need to address if they want to use a latent variable approach for psychological data. To date, researchers have not made significant progress in justifying or defining their latent variables at a psychological level (see also

Maraun, 1997; Uher, 2008a, 2008b; Uher & Visalberghi, 2016). This is an important avenue for further theory development, if we assume that there are latent variables that are meaningful and valid representations of human personality.

SUMMARY

The previous chapter cast a large shadow over the model that I proposed in Chapter 6. There is substantive variability in the internal structure and reliabilities of both traits and values, using both established instruments and lexical approaches. In this chapter, I examined whether there are some systematic processes underlying this variation. The postmaterialism hypothesis first proposed by Ronald Inglehart offers an interesting perspective to start to unravel potential mechanisms that lead to this instability. The resources available to individuals in specific populations and the characteristics of the environment in which individuals interact are important forces that shape how values and traits are elaborated and expressed. This ties directly back in with the neural network models that I discussed in Chapter 6. The two-dimensional structure proposed a pre-dispositional blueprint that is expressed differently, depending on the environment. I used a city metaphor to express this plasticity across different populations. If we conditionally accept this metaphor as a theoretical starting point, there are a number of potentially far-reaching implications. I alluded to two research implications (how to unpackage resources and the situation, and the status of latent variables in research) and I will return to the practical implications in Chapter 12. But before we get there, I need to look more carefully at these theories that have predicted different levels of traits and values across environmental contexts. This provides the other side of the puzzle, bringing together the cultural variability in traits and values that I discussed in Chapter 5 with the arguments put forward in this chapter. This is the focus and challenge of the next chapter.

9

Values and Traits as Adaptive Strategies

In the previous chapter, we saw how situational characteristics and the resources that individuals possess have a profound influence on the organization of traits and values. In the current chapter, I return to the situational context and address the question of how these characteristics may influence the importance and prevalence of specific values and traits versus others. In Chapter 3, we saw that both values and personality traits vary across cultures. What might account for these differences in relative importance and prevalence of traits and values? Returning to the personality as city metaphor used, in this chapter I discuss the importance of different city neighborhoods. My main argument is that traits and values vary across societies because they are adaptive and respond to environmental conditions.

However, rather than discussing specific individual traits and values, in this chapter I make use of their presumed underlying organizing system. In Chapter 5, I proposed an organization of both values and traits along two independently working systems: a behavioral approach-avoidance system and a modes of cooperation system. These two major systems should interact with the environment in calibrating trait and value levels. Therefore, I first review different theoretical approaches related to resources, affordances, and threats that may up- or down-regulate these underlying systems and apply these insights to the behavioral motivation by modes of cooperation model. After outlining the major arguments, I review some empirical evidence that can shed some light on the plausibility of these processes. What situational features are most salient for explaining variability in human traits and values in contemporary societies? How did traits and values adapt to various types of contextual variables? To discuss these variables, I need to draw on various theories in economics, sociology, psychology, and biology. Since the argument relies heavily on the concept of adaptiveness, toward the end of the chapter I return to the question of genes. Do genetic differences interact with these environmental conditions in predicting differences in traits and values between people living in different parts of the world? These processes have been discussed both from a gene-culture coevolution and a gene-environment interaction perspective. This last section begins to link all the parts of the book together by considering the interplay among all the biological, ecological, social, and economic variables at the different levels in predicting human personality variation. I also highlight the effects of the situation discussed in the last two chapters when looking at all these variables together. However, one step at a time – let us start by examining sociological theories focusing on the crucial role of money and wealth.

ECONOMIC RESOURCES

In the previous chapter, I discussed postmaterialism theory. As described earlier, the theory was initially proposed to account for value shifts from materialism toward postmaterialism in populations with increasing wealth. Chris Welzel and Ronald Inglehart (2010) further developed this reasoning into a universal model of human development that describes the transition from traditional agricultural economies to modern-day information societies. They proposed a four-stage process that links resources to well-being within societies. Because the first two stages are important for our current discussion, I briefly describe them here. Environment-utility linkages are the first stage in their sequential framework. The key driver at this stage is economic development. In less economically developed contexts, life is shorter and harder, with fewer opportunities to thrive. There are nearly constant threats to basic physical survival, creating existential anxieties and pressures. As a consequence, individuals in traditional agricultural economies are strongly focused on preventing further loss and prefer an authoritarian structure[1] that restricts disorder and allows some sense of stability (even though this stability is only perceived as decreasing uncertainty without providing opportunities for thriving). With economic advancement, in contrast, existential conditions improve, offering

[1] Peter Turchin (2015) discusses how power relationships have changed from more egalitarian structures among hunter-gatherers to more hierarchical and despotic regimes during the transition from hunter-gathering lifestyles to agricultural subsistence mode. These hypotheses are fascinating and await further testing in relation to the value and personality traits that are the focus of my book.

individuals longer lives in a more stimulating and challenging environment, which in turn provides many different opportunities to thrive. To take advantage of these conditions, individuals need to approach opportunities that arise and take on more agentic orientations. Hence, behavioral approach motivations become more prevalent.

As discussed in detail by Welzel and Inglehart (2010), economic development is dependent on increasing the knowledge and agency of individuals. This growth requires the activation and mobilization of the cognitive potential of the population and of the free flow of information between individuals. Improving economic conditions remove some of the existential threats that create a need to engage with others on an instrumental basis; the greater equality, autonomy, and tolerance that are brought about by those improvements also allow more trust-based interpersonal interactions. Therefore, a secondary shift encapsulated in the utility-value link (their second stage) entails greater emancipation from authority and trust-based social cooperation. In contrast, in more existential threat conditions, cooperation is driven by meeting basic survival needs. Consequently, cooperation is primarily motivated by instrumental concerns.

These processes operate at the population level, most likely playing out across generations. This is a long-term process model, because economic conditions early on in development are important for setting the parameters that drive and characterize an individual. Once set in place, there is some, but relatively less flexibility in shifting basic value orientations. Decreasing economic conditions within societies can lead to value shifts in populations (Inglehart & Baker, 2000), but the key driver of population-level dynamics captured in this model is generational change (Welzel, 2013).

At the individual level, having more income may not necessarily be associated with the same processes. Plenty of research has demonstrated that individuals with higher income are more agentic and approach opportunities, but they are also more likely to engage in cooperative behaviors that are instrumental, rather than trust based. We are probably all familiar with individuals who are well off and do not care much about the plight of others. In the literature, such associations have been demonstrated in multiple ways. For example, higher-income individuals are typically less empathetic, less prosocial, less community focused when facing uncertainty, and more likely to cheat in interactions to their own advantage (John-Henderson, Stellar, Mendoza-Denton, & Francis, 2015; Kraus & Callaghan, 2016; Kraus, Piff, & Mendler, 2009; Kraus, Piff, et al., 2012; Piff et al., 2010; Piff, Stancato, Côté, et al., 2012; Piff, Stancato, Martinez, et al., 2012; Stellar, Manzo, Kraus, & Keltner, 2011; Stephens, Markus, & Phillips, 2013). In contrast, research suggests that lower-income individuals typically show more empathy to the suffering of others and are less likely to cheat in social interactions to gain unfair advantages. Hence, there is a clear disjunction between the individual and societal level in the type of cooperation that is activated with increasing economic resources. This is a powerful reminder that contextual effects at the population level may not generalize to individual-level effects and may even be reversed.[2]

In summary, economic resources are theoretically important for the activation of behavioral approach versus avoidance motivations, which then are likely to influence the relative importance of compatible values and traits. At the population level, increasing economic resources also lead to a shift toward trust-based cooperation. At the individual level, these shifts may be more complex.

RESOURCE-DEMAND MODELS

A fundamental psychological principle is that demands and resources interact with each other (Lazarus & Folkman, 1984). In the previous chapter, several of my arguments implied such an interaction between situational demands and available resources in the cognitive differentiation of traits and values. Now, I am concerned with the relative importance of values and traits in different environments. High levels of environmental demands

[2] Another important concern with the model outlined by Welzel and Inglehart (2010) in terms of the current discussion is the relative causal ordering of mechanisms. A third stage in their sequential framework links values to life strategies. The argument is that, once certain values are activated, individuals are motivated to pursue value-congruent strategies. In their theoretical model, Welzel and Inglehart specifically discussed prevention and promotion strategies, building on work by Higgins (1998). These strategies are then operationalized as cognitive mobilization, which emphasizes informational resources and intellectual skills (e.g., in their analysis they used a knowledge index that captures enrollment in higher education, number of patents submitted per capita, and internet hosts per capita). At the operational level, there is no conflict with my arguments because their treatment of promotion and prevention strategies is quite different from mine. On the other hand, at the theoretical level, promotion and prevention strategies are conceptually close to behavioral approach and avoidance motivation. Put simply, the theoretical link between motivations and values in their framework is causally reversed compared to my proposed causal relation. I argued in Chapter 6 that both values and traits are organized and derived from basic biological systems. Behavioral approach and avoidance motivation are parts of a basic motivational system that can also be identified in animals, and this motivational distinction has links to neurophysiological and neurocognitive processes. Taking into account the preservation of these motivational systems across species and the emerging evidence of the involvement of fundamental brain networks, it is more likely that such fundamental approach versus avoidance motivational systems give rise to personality traits and values, rather than the other way around.

are detrimental for psychological functioning and become a stressor, unless they can be met by sufficient resources that offset the demands imposed on the organism (or population).

One theory that keys in on this interaction to explain value differences across populations is Evert Van de Vliert's climatoeconomic theory of culture (Van de Vliert, 2009, 2013a,b). Van de Vliert proposes that climatic demands (stressor) and wealth (resource) interact, and their combined force affects a host of psychological processes. An evolutionary starting point is that there is an optimum temperature for human survival that approximates 22 degrees Celsius on average. With average temperatures in the low 20s, humans are neither overly hot nor cold, which in turn creates less need for thermal regulation (either shivering to increase body heat or sweating to decrease body heat). In this temperature range, growth of plants is not restricted to a single harvesting season, and as a consequence, high-quality caloric food is more readily available throughout the year. In addition, there are fewer threats to health and well-being caused by either extreme cold or heat.

If the average temperature deviates from this optimum in either winter or summer, the resource needs increase to meet the demands imposed by particularly colder but also hotter climates. If these climatic demands cannot be adequately met by available resources, humans are forced to survive in suboptimal conditions, leading to a shift toward greater behavioral avoidance motivation and hence increased concerns with security and basic physical survival. Similar to the arguments introduced earlier, cooperation should also become more centered around instrumental concerns, and trust-based cooperation decreases. In contrast, when climatic demands can be met by sufficient resources, individuals and communities can relatively easily address climate-induced threats (thermal regulation, access to food and shelter). Threats become opportunities for growth, especially if those threats have been controlled by human action, and the experience thereby increases a sense of self-efficacy and individual agency. Consequently, in these ecological conditions we see a shift toward behavioral approach motivation. Exploring new situations is perceived as challenging and rewarding. If sufficient resources are available to meet existential demands imposed by climate, the modus of cooperation should shift toward more trust-based interactions and diminished instrumentally driven cooperation.

To date, Van de Vliert (2009, 2013a,b) has focused on money as the most important resource available to humans that they can use to address climatic challenges. Money is obviously a cultural invention that allows individuals to purchase additional resources to address climatic demands (warm clothes, well-insulated houses, efficient heating or cooling systems, high-quality food, etc.). The extent to which money is always the best resource to shift behavioral motivation is something that I discuss in more detail later.

PARASITE STRESS

One of the strongest pressures on any biological system is the presence of pathogens. Infectious diseases are a powerful threat to reproductive fitness and the survival of organisms. Some biologists (Hamilton & Zuk, 1982; Zuk, 1992) have even suggested that sexual reproduction might have evolved as a protective mechanism in the battle with pathogens, because sexual selection increases genetic and phenotypic diversity in a population, which hinders the spread of pathogens. The most direct biological pathogen defense mechanism is the immune system (Fincher & Thornhill, 2012; Fincher, Thornhill, Murray, & Schaller, 2008; Schaller & Murray, 2008, 2011; Thornhill & Fincher, 2014). This system involves a large number of highly specialized biological systems that help identify and neutralize intruders within the body. It is nevertheless a highly costly system to maintain and is not infallible. Launching an immune response requires metabolic energy that is then not available for other fitness-relevant tasks and is potentially dangerous for the body (e.g., it may decrease the overall fitness of the system; think if you had a bad fever and needed to react quickly to an external threat; e.g., how quick could you react in this condition if a lion was charging toward you?). When overreacting, immune responses may also destroy healthy cells and interrupt vital body processes.

At the same time, many pathogens and their signals are unknown to the immune system, which means that it only reacts once signals of an infection are clearly visible and considerable damage has been done by viruses or bacteria. Therefore, any other systems that can help the body avoid contact with potential pathogens should be highly beneficial for the survival of an organism. This is how behavioral differences start to play a vital role. In the presence of parasites, it is beneficial for organisms to eschew new and untested practices or behaviors and to avoid contact with other organisms that could harbor potentially lethal diseases and parasites. Therefore, social contact with strangers (who could carry new diseases) or engaging with novel environments or tasks (that may increase the risk of infection) is risky and therefore better avoided.

However, such a behavioral avoidance motivation is not uniformly adaptive. The distribution of parasites varies around the globe, with some environments, especially those that are hot and wet, harboring greater numbers of parasites and also more lethal ones compared to other environments. Therefore, in some environments there is a premium for avoiding potentially dangerous situations. Yet this mechanism only explains a main effect of parasite stress on behavior across different environments. For such evolutionary mechanisms to work, there needs to be variability in behavioral dispositions across individuals even within the same environment. This variability is also linked to individuals' differential susceptibility to diseases because of different physical constitutions (some people are more physically robust and healthy than others), which in turn leads to different reactions toward potential situations that may cause injury or infection. In summary, behavioral tendencies to avoid potential infection vary across individuals and environments.

The straightforward implication arising from this argument is that behavioral avoidance motivation is activated more in contexts of high parasite prevalence, whereas in environments of lower parasite prevalence, the behavioral avoidance system is less likely to be active. Hence, parasite stress theory emphasizes environmental pressure as a major causal factor in the emergence of value and trait differences, this time in the form of the prevalence of communicable diseases and parasites in the environment. The greater presence of parasites and diseases increases the risk of illness and death, which then activates the behavioral avoidance system to mitigate these risks.

EMPIRICAL TESTS OF THESE RELATIONSHIPS

In this chapter I introduced three major classes of threats and resources: economic, climatic, and biological (parasitic). How well do the different theoretical predictions of these classes stack up in empirical tests? Are some types of threat or resources more important than others? A number of studies have tested the importance of individual set of predictors, typically showing relative good support for the theoretical models (Fincher & Thornhill, 2014; Schaller & Murray, 2011; Van de Vliert, 2013a,b; Welzel, 2013).

There are nevertheless several issues with these tests that require further empirical investigation. First, many studies only focus on a single set of predictors, without testing alternative explanations. In the next section, I present one analysis in which I comparatively test these various explanations at the societal level. Yet, another important problem is that many studies focus exclusively on relationships at the aggregate level. For example, what is the relative effect of the gross domestic product per capita as a statistical estimate of wealth resources available to the average citizen of a country on the averaged personality or value mean of a sample of (typically) students? Many studies do not deal with the specific effects of wealth on the traits and values of individuals, but rather only on the aggregated levels of values and traits. More appropriate tests would involve multilevel models that directly text the context effects of threats and resources on the responses of individuals. Third, even in such multilevel tests, the threats and resources are typically statistical averages. For example, to what extent is parasite stress estimated at the country level actually representative of the living environment of any given individual? It would be more informative to obtain estimates of environmental threats and resources for each individual, instead of relying on statistical aggregates that may or may not be relevant for any given individual in a sample. There is currently relatively little information available to address these questions. Let me now review several studies and present some new data and results that provide first steps in answering these questions.

Testing the Theories at the Country Level

A first step is to comparatively test the various theoretical approaches with aggregate nation level data. As just outlined, this is not the best test, but it is a reasonable start. Therefore, for traits I combined nation level scores from three sources: the NEO-PI-R (Hofstede & McCrae, 2004; McCrae, 2001) the BFI using student samples (Schmitt et al., 2007), and samples of employees who responded to the Occupational Personality Questionnaire (OPQ) – a commercial personality test – when applying for a job (Bartram, 2013; Bartram & Brown, 2005). The scores thus came from three different instruments, using different samples, testing environments, and time frames. The combined score should provide a general overview of prevailing trait orientations in the general population that would be relatively independent of method and sample issues. However, as I showed in Chapter 4, independent measures of personality often do not correlate that highly with each other, even though they are supposed to measure the same underlying trait. The average correlations across the three different instruments were 0.64 for Neuroticism, 0.57 for Extraversion, 0.43 for Openness, 0.23 for Conscientiousness, and 0.00 for Agreeableness. The societal-level scores for Conscientiousness and Agreeableness when measured with the OPQ correlated negatively with the scores from the other two measures. The OPQ used a different format that did not involve ratings and relied on employee samples' (instead of students) scores collected in real-life testing situations. It is difficult to say what the most likely explanations of these unexpected patterns might be. As a consequence, I present the results using the average scores across all three instruments (a combined sample of 69 societies), but we need to be careful in interpreting these patterns. This is particularly the case for the Agreeableness and Conscientiousness data (I also report the individual instrument results, where they are relevant).

For values, I combined the student and teacher scores (Vauclair & Fischer, 2011) and created the higher-order major dimensions that I have used throughout the book. Data from 67 countries were available. Given that I have only a single instrument and two different types of samples, the values scores are less diverse than the personality scores. The combined value and trait scores are the dependent variables in this analysis.

To test the various different theoretical models, I used widely used indicators from previous studies. To test the importance of parasite stress, I averaged two historical disease indexes (Murray & Schaller, 2009) that capture the historical prevalence of leishmaniasis, schistosomes, trypanosomes, leprosy, malaria, typhus, filariae, dengue, and tuberculosis. To test the effect of wealth, I combined estimates of gross domestic product per capita for the years 1985, 1990, 1995, 2000, and 2005. These years capture historical estimates of national wealth that probably well characterize the national environment during the periods when the data were collected. To test climatic demand effects, I used Van de Vliert's (2009) standard index, which is computed as the sum of the deviations from 22°C (ca. 72F) for the lowest and highest temperatures in the coldest month and for the lowest and highest temperatures in the hottest month. One issue with any single index is that, especially for large countries, one score may misrepresent local populations. For example, countries like Russia, China, or the United States cover multiple latitudes and economic zones. This is an obvious limitation for the representativeness of the three major indicators that I use to test the theoretical propositions here. Van de Vliert (2009) reported some analyses showing that, (at least) for climatic effects, not adjusting for local climatic variation will result in more conservative estimates. As a result, the current analyses can be taken as lower estimates of the true effects.

I ran multiple moderated regression to test the independent contributions of each theoretical model. To test the climate-economic model, I examined the interaction between mean-centered wealth and climatic demand. Parasite stress was entered without further centering.

Table 9.1 shows the direction of the significant effects for each predictor, as well as the overall amount of explained variance. The first consistent result is that, when controlling for wealth and climatic effects, the effect of disease stress is not significant in any of the models. Despite some

TABLE 9.1 *Results of the Linear Model Analyses Predicting Cross-Cultural Variability in Traits and Values*

	E	O	A	C	N	Openness	Self-Transcendence	Self-Enhancement	Conservatism
Disease stress	n.s.	n.s.	n.s.	n.s.	n.s.	n.s.	n.s.	n.s.	n.s.
Climatic demands	n.s.	n.s.	negative	n.s.	n.s.	n.s.	n.s.	n.s.	negative
Wealth	negative	n.s.	n.s.	negative	positive	positive	positive	negative	negative
Wealth x Climatic demands	positive	n.s.	positive	positive	positive	n.s.	positive	n.s.	n.s.
Explained variance	24.3	10.8	22.5	22.9	23.6	52.7	40.1	40.8	62.9

Note: The table reports the direction of significant effects and the overall explained variance for each linear model at the societal level.

previous research showing effects on traits and values (Fincher et al., 2008; Murray, Trudeau, & Schaller, 2011; Schaller & Murray, 2008, 2011; Thornhill & Fincher, 2014; Thornhill, Fincher, Murray, & Schaller, 2010), in a direct comparative test of the various models, the independent effect of parasite disappears.

Second, wealth has a significant effect in seven of the nine models. The interpretation of wealth effects has to be done in light of its interaction with climatic demands, so I discuss it next. Climatic demands also had some direct effects on Agreeableness[3] and conservative values. Similar to wealth effects, these effects have to be interpreted in light of the interaction between wealth and climate. Focusing on these interactions, five of the nine models showed significant interactions between wealth and climate. What is noteworthy and somewhat surprising is that climate-economic effects are consistently stronger for personality traits than for value priorities. The climate-economic model has been extensively applied to value differences, but not to traits so far.

To interpret the interactions, we can plot the relative effects of wealth and climatic demands. Figure 9.1 shows a prototypical interaction. I split countries into high- and low-income countries, so that we can see the relative effect of climatic demands in those societies where individuals on average have higher incomes compared to those where individuals have lower incomes.

The interaction between wealth and climate is in line with the predictions that I presented earlier. Overall, higher wealth functions as a buffer in countering a negative effect of climate on personality traits. For example, in Figure 9.1, Extraversion is higher in demanding climates, but only when sufficient resources are available for individuals on average. In high-income countries, more climatic demands are associated with a significant increase in Extraversion, Conscientiousness, and Emotional Stability. In contrast, in high-income countries with less demanding climates, wealth does not have much of an effect on values. In contrast, in lower-income countries, climatic demands are associated with significantly lower levels of Agreeableness, Conscientiousness, Emotional Stability, and Self-Transcendence values (the latter two effects are approaching significance, $p < 0.10$). Therefore, in wealthy countries climatic demands become a challenge that can be addressed.

[3] As noted earlier, Agreeableness scores did not correlate with each other. The effect of climatic demands was significant and negative for both the BFI and NEO scores. The interaction reached significance only for BFI data, but the direction of the interaction was the same for OPQ and NEO scores.

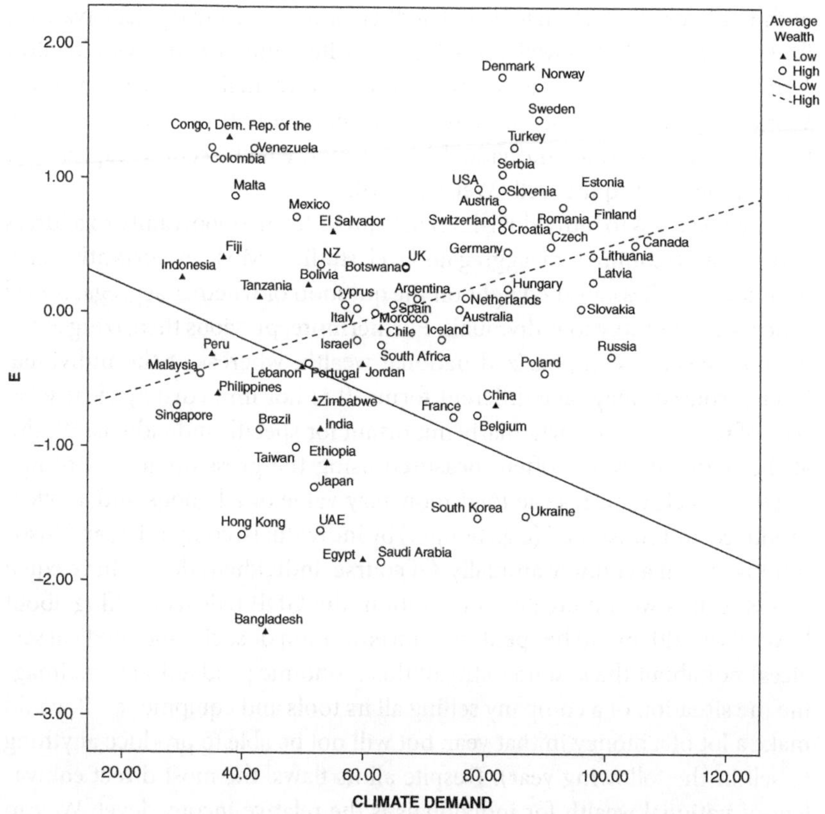

FIGURE 9.1 Climatic demands by wealth interactions on Extraversion.

Overall, the explained variance for personality was smaller than for values. Values are probably more directly related to conscious goal strategies and hence might be more sensitive to direct environmental characteristics. The most consistent influence for values was wealth. Greater wealth was associated with raising Openness and self-transcending values and diminishing Self-Enhancement and conservative values. Since wealth in modern societies is directly linked to lifestyles and the realization of possibilities in people's lives, direct effects might well be stronger for values than for personality.

A Multilevel Test Focusing on Values of Individuals

The test in the preceding section focused on nation-level associations. To what extent do these processes influence the personality and value levels of individuals? Unfortunately, to date there are no secondary data available for traits that include a well-validated multi-item measure that would allow a sufficient test of these associations at the individual level. However, the World Value Survey that I have relied on repeatedly throughout this book has short measures for individual-level values, which seem to capture the main dimensions quite well (Welzel, 2010).

Having access to individual-level data provides an opportunity to address various shortcomings of aggregate-level studies. Most importantly, such multilevel studies can shed light on the question of whether aggregate-level processes also apply to individuals. Furthermore, previous theorizing at the national level has emphasized national wealth, whereas at the individual level, resources may take different forms. It is not immediately clear what type of resources are particularly important for specific individuals. Wealth at the national level is often measured using the gross domestic product (GDP), which captures the total monetary value of all goods and services produced and consumed (e.g., bought) by individuals, companies, and institutions within a country annually. Of course, individuals do not have equal access to this wealth created; in addition, the GDP tells us nothing about how this wealth might be spent (e.g., on education or social and medical services) nor about the sustainability of this economic production (e.g., imagine the situation of a company selling all its tools and equipment – it would make a lot of a money in that year, but will not be able to produce anything to sell in the following year). Despite all its flaws, the most direct equivalent of national wealth for individuals is the relative income level. We can examine whether economic resources at the national level (GDP adjusted for population size) and resources at the individual level (income of the individual) show independent additive effects on the endorsement of values. Which has a greater impact on the values endorsed by individuals: their personal resources or the general wealth in a nation that influences institutions and societal processes? As I have already mentioned, these effects may not always work in the same direction.

What other resources could be relevant? For example, money may not be a good indicator of actual capacities that an individual possesses to address environmental demands. As mentioned earlier, Welzel (2013) proposed that cognitive resources are important for human empowerment. The most direct way to gauge cognitive resources of individuals in broad surveys is to measure their completed level of education. Individuals who have higher education are likely to be better equipped to deal with challenges in their environment and to develop strategies to overcome these challenges and turn them into opportunities for growth. Only individuals who face

TABLE 9.2 *Overall Results Testing the Relative Effects of Economic Resources, Parasite Stress, and Climatic Effects on Values of Individuals*

	Openness to Change (vs. Conservatism)	Self-Transcendence (vs. Self-Enhancement)
Economic effects		
Education	Positive (replicated)	Positive (replicated)
Personal Income	Positive (replicated)	Negative (replicated)
National Wealth	No effect	Positive (replicated)
Parasite Stress Model		
Parasite Stress	No effect	No effect
Climato-economic Model		
Climate x National Wealth Interaction	No effect	No effect
Climate x Personal Income Interaction	No effect	Significant in only one study, not replicated
Climate x Education Income	Significant (replicated)	No effect

manageable challenges in their environment will be set on a path to develop greater Self-Transcendence and Openness values. In short, educational level is a likely individual-level resource over and above personal income that is worth exploring in more detail when discussing both human empowerment and climate-economic theory effects.

I conducted a study with World Value Survey data that explored these effects. The first wave of the survey in which data on the circular structure by Schwartz were available was conducted between 2005 and 2009 and included representative samples from 56 countries (N = 71,916). The next wave with available data was conducted between 2010 and 2014 and included representative samples from 52 countries (N = 74,042). In both waves, respondents answered short items that were modeled on the PVQ; therefore, those items were more similar to trait items. I created difference scores that capture the overall value dimensions underlying the circular structure (see Figure 6.3).

Thus the same instrument was administered to large representative samples in two independent studies. This provides an exciting and unique opportunity to test the replicability of the effects.

How well do the various theories work when tested on values of individuals directly? Table 9.2 shows the overall results. In the data collected between 2005 and 2009, national wealth did not significantly predict Openness to change (versus Conservatism) values. However, greater income and education of individuals within nations were significantly associated with endorsement of Openness values, in line with my overall predictions. Individuals with higher relative income and more education endorsed more Openness values. In line with findings at the national level that I presented earlier, parasite stress was not significant when predicting values of individuals. Neither was the interaction between climatic demands and national wealth. However, when focusing on individual resources, education level made a difference depending on the type of climate to which an individual was exposed. There was a weaker effect of education on Openness values in benign climates, but the effect became much stronger in demanding climates. Examined the other way around, for individuals with university education, climatic demands were not a statistically significant predictor. Climate was no longer a stressor for educated people. For individuals with no schooling on the other hand, climatic demands made a significant difference, and greater climatic challenges were associated with more conservative and less open values. However, it is important to note that all these effects at the individual level were relatively weak, explaining only about 3% of variability between individuals. The national-level effects explained about 25% of variability between national samples (remember that this amounts to about one-quarter of the about 12% variance that is due to societal differences, see Chapter 5).

For Self-Transcendence versus Self-Enhancement values, wealth effects were in opposite directions at the individual and national levels. In higher-income countries individuals reported higher levels of Self-Transcendence values overall; however, those with higher personal income reported higher Self-Enhancement values. Hence, for income the effects diverged between individuals and societies. Education levels at the individual level were associated with greater Self-Transcendence values, as would be predicted from the theoretical reasoning that I outlined earlier.

Greater parasite stress at the nation level was not significantly associated with higher levels of Self-Transcendence values. Climatic demands and wealth at the nation level again did not significantly predict values of individuals. However, climatic demands and individual-level income had a significant interaction in predicting Self-Transcendence versus Self-Enhancement values. The relationship between income and values was strongest in highly demanding climates and was significantly weaker in benign climates. As with Openness to change versus Conservatism values, individual-level effects did explain a relatively small amount of variance (about 5%), and nation-level effects explained a bit more than 25% of the variance between nations (which is about one-quarter of the average 12% due to societal differences; see Chapter 5).

The exciting finding is that these results were replicated in a new sample collected approximately five years later. For Openness to change versus Conservatism values, none of the national-level effects reached significance, similar to the previous data set. At the individual level, both income and education were significant again, now explaining slightly more than 10% of the variability between individuals. The cross-level interaction effect between climatic demands at the nation level and the education level of individuals was significant again, replicating the effects from the previous study. Climatic challenges have an effect on individuals with lower levels of education, but not for individuals who have spent more time in school and university.

Also replicated were the diverging effects of wealth at the individual versus nation level for Self-Transcendence versus Self-Enhancement values. Greater income at the individual level was associated with a higher endorsement of Self-Enhancement values, whereas individuals residing in nations with higher wealth reported endorsing greater Self-Transcendence values. None of the other nation-level effects were significant. The previously encountered interaction between individual level income and climatic demands was not replicated.

Summary of the Comparative Tests

Which variables are important for shaping traits and values of individuals (see Table 9.2)? The aggregate-level study suggested that traits are more responsive to the interplay between climatic demands and aggregate resources (wealth). For values, we saw quite strong wealth effects at the aggregate level. Yet, these effects largely disappeared when separating the effects of personal income versus national wealth in a multilevel study testing the separate effects of income on values reported by individuals. Only for Self-Transcendence versus Self-Enhancement values did national wealth matter. Individual-level resources, examined as both relative income and the level of education, do matter for value endorsements of individuals. These income effects provide support for arguments that having access to greater economic resources is associated with higher autonomous and agentic motivation of individuals. Having more wealth at one's disposal increases choices for individuals, unleashing a drive toward greater emancipation of individual desires and actions (Welzel, 2013). The interaction between education and climate was also very interesting in this context, because it was education but not personal income that was important for buffering climatic demand effects on Openness to change versus Conservatism values of individuals.

Returning to the differential effects of individual and national effects on Self-Transcendence versus Self-Enhancements effects, it is important to reemphasize that effects for individuals may not generalize to nations and vice versa. Greater wealth at a national level was associated with greater concern for others, whereas individuals with higher income reported less concern for others and more concern with one's Achievement and Power. It is likely that aggregate wealth effects bring about contextual changes that allow individuals to become less self-centered and to consider more the well-being of others close and distant to them, as well as the conservation of the natural environment. In Chapter 6, I proposed that the underlying motivation for Self-Transcendence versus Self-Enhancement values is the mode of cooperation, either focused on trust or on instrumental control. Individualized resources only activate motivations to seek out situations and opportunities to further demonstrate success and achieve power over others, with a decreased concern for the welfare of others. In contrast, if everyone has more resources available, this shifts the social climate toward greater trust, which allows individuals to shift their cooperation orientation. The larger contextual economic conditions are likely to set a social climate that influences how individuals operate and how their personality dynamics develop. This most likely involves developmental processes, which I discuss in the next chapter.

POPULATION-GENETIC DIFFERENCES: GENE-CULTURE COEVOLUTION

So far I have focused on situational demands, threats, and resources. When discussing evolutionary adaptation, a biologist might also consider the possibility of genetic changes that help an organism adapt to different environments. In Chapters 5 and 6, I discussed heritability estimates (implying genetic differences) as well as specific neurotransmitter systems that are regulated by genes that vary between individuals. To briefly recap, there is considerable evidence that dopaminergic and serotonergic brain systems regulate basic approach and avoidance motivations that then give rise to traits and value differences between individuals. However, an important twist to this individual difference story is that these genes do not only differ between individuals: we now have evidence that distributions of these genes also vary to some extent between populations. This raises not only a considerable number of ethical issues but also many theoretical questions. Right now, my focus is the extent to which these population genetic differences may explain differences in personality traits and values across the world.

The most famous gene to have been studied in this regard is the 5HTT gene on chromosome 17, which is involved in the regulation of serotonin. Two major variants have been described – a short allele (S-allele) and a long allele (L-allele) – that result in different levels of serotonin expression in the brain (Hariri, 2009; Lesch et al., 1996). Individuals with the short allele are thought to have higher serotonin concentrations in the synaptic cleft after release and therefore less serotonin available for functional purposes, compared to those with a long allele. Joan Chiao and Katherine Blizinsky (2010) were the first to relate such systematic variation in the global distribution of a functional variant of this gene to cultural differences. They found that the distribution of the gene around the globe correlated with socially oriented (collectivism) and hierarchy-oriented values, which in my terminology correspond to Conservatism and Self-Enhancement values. Chiao and Blizinsky interpreted this pattern in light of a gene-culture coevolutionary process. As discussed in Chapter 6, serotonin is a major contributor to the behavioral avoidance system. Individuals with at least one short allele of this gene are more likely to focus on negative emotions, avoid risks, and engage in more narrow thinking, whereas individuals with two long alleles are more likely to focus on positive emotions and are willing to take risks. Chiao and Blizinsky argued that environments with a greater presence of pathogens conferred an advantage on S-allele carriers: their greater avoidance of risks led to an increased likelihood that they would survive and procreate in environments that were rife with pathogens. Hence, the environmental pressures led to the selection of a phenotype that was more risk averse and group oriented.

I am highly sympathetic to the meta-theoretical approach of gene-culture coevolution. At the same time, the available evidence is not convincing enough at this stage to make such long-reaching historical claims. After all, it is based on a limited number of contemporary populations. We do not have any evidence so far on historical changes in serotonin genes as populations moved around the globe and encountered different levels of parasite stress. The emerging field of paleo-genetics is likely to shed exciting light on these possible genetic changes in the near future. This is an important avenue of scientific inquiry to watch, and I will return to the opportunities and challenges of gene-culture coevolution processes in the final chapter.

Other problems with the current approach have been raised, most notably questions about the extent to which selection pressures actually have caused such changes in the genetic code or whether neutral processes such as genetic drift or founder effects (see Chapters 2 and 11) might be more parsimonious and accurate descriptions of these differences. Yet, independent of the causes of these differences, given that these genetic variations exist across populations around the world, we have to consider whether such population genetic differences exert some influence on expressions of traits and values. As I outline in the next section, the most plausible process is the presence of gene-environment interactions. In other words, population-level differences in genes might matter in certain environments, but not in others.

POPULATION GENETIC DIFFERENCES: GENE-ENVIRONMENT INTERACTIONS

Both serotonin and dopamine are important for regulating processes that give rise to personality differences between individuals (and potentially different groups of individuals). The same individuals or groups of individuals may show quite different behavioral reactions if placed in different environments that require different behavioral approaches to ensure survival. Some environments are more threatening, requiring cautious and careful exploration when finding food and attracting mates; other environments are more benign and less threatening, so too much caution is counterproductive because it may lead to missing resources valuable for survival and chances for attracting a high-quality mating partner. In short, genetic predispositions are likely to interact with environmental conditions. There is initial evidence that this is indeed the case.

For example, taking the same serotonin database created by Chiao and Blizinsky, I examined whether environmental stressors lead to different value orientations depending on population-level genetic predispositions (Fischer, 2013a). My assumption was that populations with more people who are susceptible to stress (higher frequency of S-alleles) should be more vulnerable in threatening environments. Individuals with greater levels of depression and anxiety might be less likely to survive because they do not extract sufficient resources from the environment to survive and may not form strong alliances to deter competitors. One plausible protection mechanism in such a scenario is to endorse more hierarchical and conservative values. This insight has been derived from historical analyses of social groups: hierarchical groups have been shown to be more successful in warding off threats and surviving longer (Sosis, 2000; Sosis & Bressler, 2003), which in turn lead to increasing socialization pressures toward endorsement of hierarchical and conservative values. I therefore proposed that endorsing hierarchical and conservative values within social groups that were likely to be susceptible to greater stress reactions would lead to greater social stability because hierarchical values (and organization) would allow these groups to be more efficient in providing protection against environmental threats and stressors.

These predictions were generally supported when testing a variety of different environmental stressors, including parasite stress, climatic stressors, and threats of violence. Therefore, this was a first proof-of-concept study that suggested that gene-environment interactions across populations today may influence value orientations.

In a follow-up study, my colleagues and I focused on dopamine (Fischer, Lee, & Verzeijden, 2017). We examined genes that have been shown to influence dopamine processes at the neural level in previous individual-level studies. We then searched for studies that had examined the distribution of these genes around the world. After reviewing many studies, we included seven genes that had been studied in at least 10 different populations in previous research. These genes did covary together, suggesting that populations that had one genetic variant associated with a higher or lower efficiency in dopamine processing in the brain also tended to have other genetic variants in other genes that were similarly effective. As I discussed earlier, it is difficult to interpret such patterns because these differences could have arisen as a functional adaptation to local environments or be due to random changes. However, what we can investigate beyond doubt is whether such population genetic differences interact with environmental conditions. In our study we focused on climatic demands in predicting personality differences across populations. Climatic demands strongly affect core survival needs of humans, including the need to maintain a stable temperature and find sufficient food for survival; if the climate is extreme and restricts the growth of both plants and animals, it will be more difficult to find food.

Our main assumption was that both behavioral approach and avoidance strategies might be adaptive in high-climate-stress environments, depending on the reward parameters of the biological system. For example, approach behavior is likely to yield additional resources through the exploration of novel situations, but it also heightens risk due to increases in the number of violent encounters with other human groups and predators and of the incidence of diseases and injuries. In contrast, avoidance behavior may reduce overall risk, yet it also decreases the chances of obtaining additional resources that might increase fitness. Because dopamine is centrally involved in reward processing, we speculated that genetic variation in dopamine production or sensitivity may mediate the personality traits a person might develop. More specifically, we examined Extraversion and Neuroticism because these two personality traits might be particularly strongly related to dopamine processes.

This thinking draws heavily on work by Daniel Nettle (2006), which I discussed in the previous chapter and will discuss in more detail in Chapter 11. As discussed previously, highly functional dopaminergic systems are likely to be associated with Extraversion and Emotional stability, while less functional dopaminergic systems are more associated with Introversion and Neuroticism (Read et al., 2010). The key thought is that a person with a highly functional dopaminergic system may develop a different personality profile under a stressful compared to benign environment. In highly demanding environments, individuals with a highly functional dopaminergic system are more strongly and positively reinforced by rewards than by the negative experience of risk. Thus, we proposed that they will develop behavioral tendencies toward stronger approach motivation (higher Extraversion and lower Neuroticism). On the other hand, in high-stress environments, individuals with less efficient dopaminergic systems will shift toward avoidance behaviors, because risk is perceived as higher and more aversive. The contrast condition is the benign environment that is most suitable for human thriving – here resources are more abundant (decreasing the reward incentive for approach-motivated individuals), and risks are less severe (decreasing the risk component for avoidance-motivated individuals), therefore reducing the motivational differences between the groups differing in genetic predisposition.

In summary, we expected greater differentiation of personality phenotypes in high-stress environments, due to the differential reward versus risk sensitivity associated with the personality phenotypes, compared to environments with relatively benign climates (where we did not expect strong personality differences).

We created a genetic index for dopamine efficiency drawing on 807 independent samples involving 127,685 participants across 73 societies or territories. We then examined how this index interacted with climatic demands in predicting three measures of both Extraversion and Neuroticism. Overall, our predictions were supported. The interaction on average explained another 12% of variability in personality scores over and above the main effects of genes and climate, and after controlling for both wealth and parasite stress as alternative explanations of any societal differences. Figures 9.2

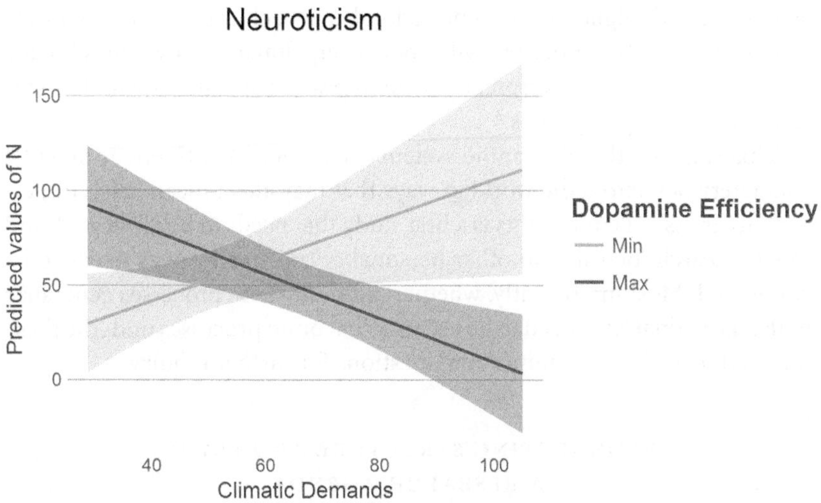

FIGURE 9.2 Interaction between dopamine genes and climatic demands for predicting overall Neuroticism scores.

and 9.3 show the interaction when combining all the personality scores into overall Extraversion and Neuroticism trait scores at the societal level. As can be seen, genetic predispositions diverge most in highly demanding climates. Highly functional dopaminergic systems at the population level are

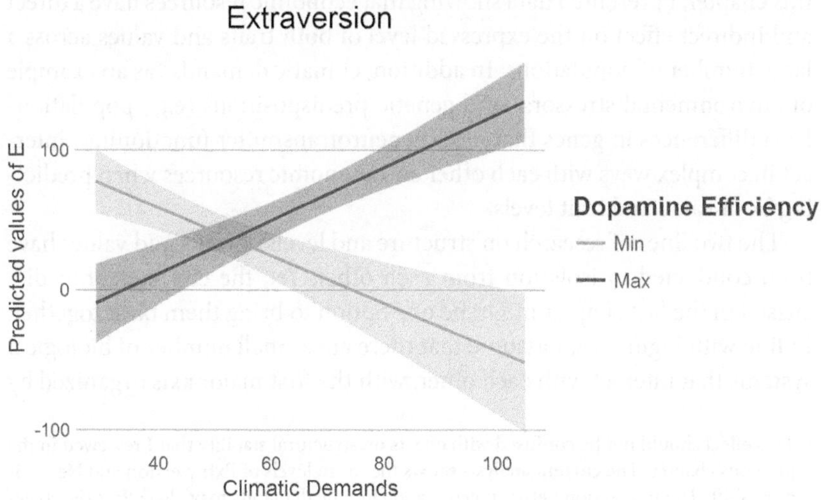

FIGURE 9.3 Interaction between dopamine genes and climatic demands for predicting overall Extraversion scores.

associated with higher levels of behavioral approach traits (more Extraversion and lower Neuroticism) with increasing climatic stress, but climatic demands do not affect personality phenotypes in populations with less efficient dopaminergic systems.[4]

This suggests that dopamine systems also interact with environmental characteristics across the world in ways that may affect mean levels of personality traits. Of course, this is a first study that needs to be replicated, and more research focusing on other personality traits and values needs to be conducted. Most importantly, whether and how these processes generalize to the individual level need to be tested. What brain processes underlie these interactions? These are important questions for further inquiry.

RECONNECTING STRUCTURE AND LEVELS: A RESEARCH AGENDA

In the previous chapter, I reviewed evidence suggesting that (a) values become more differentiated with increasing economic development, (b) Big Five traits become more reliable and internally valid with increasing national wealth, and (c) Big Five traits and values become more strongly connected with increasing resources and reduced environmental threats. There was also some evidence that traits might be differently structured (showing a different structure or the emergence of additional traits) in environments that put different economic and social demands on individuals. In this chapter, I presented data showing that economic resources have a direct and indirect effect on the expressed level of both traits and values across a large number of populations. In addition, climatic demands (as an example of environmental stressors) and genetic predispositions (e.g., population-level differences in genes that modify neurotransmitter functioning) interact in complex ways with each other and economic resources when predicting both value and trait levels.

The two lines of research on structure and levels of traits and values have been conducted in isolation from each other. Yet, the city metaphor discussed in the last chapter might be one option to bring them back together. In line with Figure 6.3, I assume that there are a small number of biological systems that interact with each other, with the first major axis organized by

[4] This effect should not be confused with effects on structural stability that I reviewed in the previous chapter. The current analysis shows the mean levels of Extraversion and Neuroticism shift. The interaction between genes and climatic demands may also affect the structural stability of the two traits, but we were unable to test this.

behavioral approach and behavioral avoidance motivation (which are two independent systems arranged on a single axis) and the second axis being formed by different modes of cooperation. The interplay between these different biological systems gives rise to different value orientations and behavioral traits. In different environmental contexts, these behavioral approach versus avoidance systems and modes of cooperation are differentially activated, depending on the functionality of these resulting sets of behaviors. For example, in economically scarce conditions, instrumental modes of cooperation are likely to be more salient. How these modes of cooperation are expressed may depend on (a) the specific social context and (b) environmental conditions and available resources (independent of money). For example, if individuals in such an economically deprived setting are embedded in a tight social network and have to negotiate demands from others in this tight network, but can also solicit help from this network, they are likely to develop a richer set of social behaviors and values that are adaptive in this context for meeting survival and reproduction needs. These behaviors and values are salient in this context, leading both to higher prevalence (increasing their specific levels) and to a richer cognitive differentiation and elaboration (a more differentiated and elaborated structure).

In contrast, in more economically well-developed contexts a different set of traits and values will be both salient and cognitively differentiated. For example, in many Western societies individuals are only loosely integrated in a diverse network of social relations across a number of domains (family, work, sports, other leisure activities, etc.). Therefore, making contact and connecting to new individuals might be more important than negotiating and maintaining tight relationships. There is also less need to solicit financial help from others or to signal one's financial status to others because individuals on average are more financially independent. Therefore, modes of cooperation shift to being more trust based (leading to the greater importance and cognitive differentiation of Openness to change values and Openness traits). In contrast, conservative values and some aspects of Conscientiousness may become less salient (less important/less frequently expressed), and individuals may make fewer cognitive distinctions.

This argument only considers situational variables in interaction with social structure and resources. It is possible to introduce differences in the underlying biological predispositions. Doing so will, of course, lead to more complex interaction models.

I am painting a picture that is highly complex and interactive. To test these ideas, it is necessary to bring a large team of researchers together. They would be engaged in identifying individuals and populations that (a) vary in biological predispositions, (b) are located in different ecological environments, (c) have different social structures, and (d) have different resources available. There are likely to be interdependencies between these factors as well (e.g., certain social structures are more likely to emerge in certain ecological and economic conditions than others). It is not feasible to test all these interactions in a single study. Yet, it should be possible to develop a research program that can start chipping away at this goliath. In the previous two chapters, I reviewed a few studies that tried to test the interplay between limited sets of possible factors. Extending these studies would be an important step to reach a deeper understanding of the complex personality dynamics. Such studies could use techniques such as multidimensional scaling, which do not come with some of the problematic assumptions of latent variable techniques and can also provide a richer picture of the interrelatedness of theoretical constructs (for a fruitful application in the emotion domain, see Fontaine, Scherer, Roesch, & Ellsworth, 2007).

A second line of research that might be informative is comparative personality research, examining what behavioral traits emerge in what animal species and how these traits might relate to their social structure and the environment that they inhabit. For example, simple social hierarchies seem to be associated with a more limited number of behavioral traits, whereas animal species with more complex social organization seem to have personality structures that start to resemble more the prototypical Big Five (Freeman & Gosling, 2010; Gosling, 2001, 2008). Similarly, the environment (captive vs. wild) and individual life histories (reared by humans vs. reared by animal group) seem to make a difference for the potential structure and expression of personality traits (Uher, 2008; Uher & Visalberghi, 2016). Research on animal personality can provide interesting insights because of the larger variation of social structures and ecological niches that are occupied by different types of species.

SUMMARY

In this chapter, I focused on variability in trait and value levels in populations around the world, drawing on major theoretical perspectives that consider environmental threats and opportunities. In light of currently available data, the most important direct effect on both values and personality traits is wealth, both at the aggregate society level and via the income and educational resources directly available to individuals. Individuals with more income and living in richer societies tend to report values and traits that are regulated by behavioral approach motivation. Cooperation effects (the distinction between Self-Enhancement vs. Self-Transcendence values) differ between levels, suggesting that it is important to consider context effects beyond the immediate resources available to the individual.

Shifting the focus to genetic predispositions, a second major insight is that conditions in the broader environment interact with genetic predispositions. I discussed two proof-of-concept studies in which genes and the environment jointly influenced specific values and traits. Once we get a better understanding of the various neurotransmitters and neural functions involved in shaping trait and value dynamics, this will open opportunities to further investigate how predispositions interact with the context in which an individual or groups of individuals are operating. Personality is complex, and it is unlikely that simple main effect models (in which one variable strongly and reliably predicts a single trait or value) are adequate. It is more likely that various environmental conditions interact with each other, as well as with complex sets of genes that make up the genetic predispositions of individuals (and populations). The patterns that I discussed here might be just the top of the proverbial iceberg, because such more complex interaction effects require large samples of many populations that are only slowly becoming available for analyses.

If we return to Tina's and Sandra's behavior, we can now start to get a better appreciation of the various variables that might play a role. The overall differences between people around the world are relatively small, yet the variables that influence this small amount of variance are complex. Tina is embedded in a tightly knit social network. She performs a highly valued role in her social context, and many people in her extended network look up to her. Her set of social skills that form her traits and what she values as important are slightly different from the traits and values that Sandra values, even though both highly value their social relations. At the same time, Tina's economic conditions are considerably poorer than Sandra, with Tina having significantly more concerns for making daily ends meet. This allows Sandra greater freedom to pursue different values and personal interests than Tina. These different freedoms enjoyed by the two of them will lead to both slightly different structuring and importance of their traits and values. Both Tina and Sandra are accomplished and highly successful in their local environment, yet the specific set of traits and values vary because of the demands that are placed on them. At the same time, they both inherit slightly different sets of genes that drifted apart over the last 50,000 years since humans moved out of Africa. These sets of genes now also contribute to how both of them react to their quite different environments. To tease these effects apart, we need interdisciplinary research that captures these dynamics in large samples.

The current chapter provided snapshots zooming out to focus on the larger picture of how the environment and genes shape the distribution of traits and values across human populations. In the next chapter, I am going to discuss how these processes might play out over the lifespan of an individual. Therefore, I will zoom back in on the developmental sequences of how such differences emerge.

10

Traits and Values across the Lifespan

In the previous chapters, I discussed how the interplay between environmental conditions and internal dispositions influences both the organization and relative importance of traits and values. These discussions were informed by data that were collected at a specific point, forming a snapshot frozen in time showing the differences across populations around the world. However, such patterns obviously develop over time; people are not born with a fixed set of traits or values. Equally, these patterns probably have evolved over our evolutionary history since our ancestors split from other great apes. The changes over the lifespan and the associated dynamic developmental trajectories are the focus of this chapter. Many ideas will be somewhat speculative because we still have relatively little longitudinal data on how personality and values may develop over time for any single human being. Even less data are available on the historical developments across the evolution of us humans (which I tackle in the next chapter). I can only ask for some patience as I lay out the available evidence; collectively we need to do more research to investigate these dynamic processes further.

SETTING THE HUMAN LIFE STAGE: THE IMPORTANCE OF EARLY SOCIALIZATION EXPERIENCES

In the earlier discussion of contextual influences on value and personality ratings, there is one contextual influence that I have not touched on so far: the family environment. The family constitutes the immediate social environment within which a child grows up. We cannot ignore its importance because of the central influence that caretakers, especially mothers, have on infants. Our brains are too large to be fully formed in the womb; therefore, human babies are born relatively prematurely and require significant postnatal care and support. Obviously infants have various needs that must be met: for their bodies and brains to grow, they need to feed at regular intervals, and they need to be kept warm and sheltered from the natural elements and from potentially dangerous animals, parasites, or other humans. How to best satisfy these needs may depend on the local environment. Babies and young children are highly sensitive to environmental cues because they need to calibrate their expectations and strategies to satisfy crucial needs for survival and growth based on the available support and resources available to them. Is food available on request? What type of request is most efficient in eliciting food – should I cry only when really hungry because my caretakers immediately respond to it, or is it better to cry for prolonged periods of time to eventually get some food? If I am unwell or cold, what type of support is available and how fast? Maternal behavior is crucial for answering all these questions because of the dependence of human infants on (typically maternal) caretakers to provide food and care for survival.

If a baby experiences an environment in which food, warmth, and care are relatively consistently available on request, the infant is likely to develop a sense of security that allows him or her to explore the environment. If dangerous situations are encountered, these infants can expect their caretakers to step in and mitigate the risks. From our Western perspective, such babies in typically stable and supportive family environments appear well adjusted. In contrasts, if infants encounter an environment where they lack adequate support, care, and maternal warmth, they are more likely to develop behavioral patterns that might appear dysregulated at first sight, but actually can increase their survival rates in such adverse environments. For example, strategies might shift toward insecurity and dependency in exploring the environment and relating to others. Babies might be clingy or cry extensively. Infants might react impulsively, be antisocial, or be willing to take unnecessary risks (from an outsider's perspective). Such behaviors might appear dysfunctional from a perspective where support, care, and warmth are readily available, but these strategies can be highly beneficial for survival if the environment is less stable and supportive (Bakermans-Kranenburg & van Ijzendoorn, 2011; Belsky & Pluess, 2013; Ellis, Boyce, Belsky, Bakermans-Kranenburg, & van Ijzendoorn 2011).

I hope these examples make it clear that the early socialization environment is likely to calibrate the behavioral responses of an individual to the larger social and ecological environment later on in life. As a consequence, secure social environments early on in a person's life might shift strategies toward greater behavioral approach motivation and more trust-based cooperation. In contrast, in more insecure, cold, or inconsistent socialization environments, infants might shift toward more behavioral avoidance motivation (or potentially toward extreme behavioral approach motivation) and more instrumental cooperation. These processes have been investigated primarily in relation to potential problem behaviors (such as aggression, depression, anxiety, and school problems) that may be clinical endpoints of the various trait dimensions discussed in this book (Bakermans-Kranenburg & van Ijzendoorn, 2011; Belsky & Pluess, 2013; Dunkel, Mathes, Kesselring, Decker, & Kelts, 2015; Ellis et al., 2011; Jaffee et al., 2015; Nettle, Frankenhuis, & Rickard, 2013). This work is also very interesting because it links genetic predispositions with environmental conditions in shaping a child's developing personality characteristics, very much like what I argued in the later sections of the previous chapter.

These processes are also brought into sharp relief when examined against the different maternal care patterns that exist in populations around the world. Extensive work among hunter-gatherer societies has demonstrated that mothers typically respond very quickly to the crying of their babies (which they typically carry with them), babies can breastfeed whenever they are hungry, and mothers are overly indulgent and caring (Hewlett & Lamb, 2005; Konner, 2005, 2016; Kruger & Konner, 2010). These infants then grow up to become autonomous and confident children and adolescents who are eager to engage in more self-directed learning and exploration of the environment (Hewlett & Lamb, 2005; Konner, 2005, 2016; Kruger & Konner, 2010). In contrast, in some farming communities that have been more extensively studied than those hunter-gatherers, mothers are more restrictive in their provision of food and their attention to crying babies, which then is associated with more obedient and less autonomous behavior of children later on (e.g., Barr, Konner, & Bakeman, 1991; Fouts, Hewlett, & Lamb, 2012). Similarly, a large number of studies have shown that early childhood environments characterized by economic (leading to malnutrition and lack of stimulation) and social deprivation (e.g., abusive family environments, violence) are associated with poorer socioemotional and cognitive development (Walker et al., 2011). Family environments can have a more important influence than the cultural context. For example, a comparative study of Cameroonian, Chinese, and German young adults demonstrated that differences in conservative values could be best explained by the number of siblings that a respondent had (Bender & Chasiotis, 2011). These studies do not allow us to examine trait or value differences between individuals, but they do point to the importance of environmental conditions during infancy as a predictor of later developmental pathways.

Such influences and processes extend beyond the immediate family environment. Any resources that are available (or not) and threats that are present (or not) during early life of an individual are likely to influence trait and value development of the growing person (Nettle et al., 2013). The general process is often called ***predictive adaptive response*** (Nettle et al., 2013), and it can be imagined as a kind of weather forecast for the individual, although instead of predicting whether it will be sunny this afternoon, it tells you what is life going to be like when you are growing up and become an adult. The basic premise is that the predictive adaptive response tunes the system of an individual to respond appropriately to the current (and anticipated future) environment. This early life weather forecast allows an individual to prepare for life ahead by helping him or her to develop optimal strategies within a specific local environment. There are two important conditions that enable this forecast to be predictive. First, the early life experience needs to provide valid cues of what life as an adult is like – that is, the stressors and resources available during infancy and early childhood need to tell the developing individual what is important later in life. The second condition is that the environment has to be reasonably stable (e.g., correlations from year to year of about 0.95 are necessary according to some statistical simulations; Nettle et al., 2013) for adaptations of the individual to the environment to provide some advantage. For many of the resources and threats discussed in this book (e.g., wealth, climate), this level of consistency over time is probably realistic.

A second, but less frequently discussed, explanation of this early adaptive response mechanism is that early life conditions physically change the body of an individual and therefore bring about changes in the internal conditions of his or her system. Having less nutritious resources available as an infant means that less energy is available to fully develop brain tissue, muscles, organs, and so on, or less energy is available for repairing body tissue in case of injury or sickness (Walker et al., 2011). This reasoning is very similar to arguments that have been put forward for explaining parasite stress effects on the development of intelligence among children (Eppig, Fincher, & Thornhill, 2010). This general stress effect on the body then sets in motion accelerated maturation and other biological responses that allow an individual to reproduce faster. According to this interpretation, adaptiveness is primarily driven by internal changes to the body, rather than by fine-tuning behavioral responses to external environments (as in the first perspective).

These two processes, which highlight the complex interaction between the internal states of an individual's biological system and the external environment, are obviously not independent and may occur in tandem in real-world settings. These adaptive responses conditioned on the early developmental environment then explain changes in behavior that occur later in life (e.g., time of menarche, first childbirth, psychological stress responses, mortality). Personality in a broad sense is produced by the dynamic states of the neural and physiological system, involving many interacting neurotransmitters (as well as ion channels and hormones). The fine balance within these systems is likely to be affected by both internal and external events that unfold over the development of an individual. The early stages of development are particularly important for setting the core features of what will become the parameter space in which the personality of the individual can develop.

CHANGES DURING CHILDHOOD AND ADOLESCENCE

Once the basic parameters are set, they need to be tuned, but changes are obviously not going to stop. One particular time of development that has captured the attention of many scientists and laypeople is adolescence. After all, this stage is thought to be the time when individuals make the transition from being a child to becoming an adult, with all the associated uncertainties that come with testing out new identities and growing into a new and different role (Erikson & Erikson, 1998; Marcia, 1966). Adolescence is the formative period of an individual's life, with adolescents learning about themselves and their place in life as they actively create a new identity and develop into a full person.

There have been a few longitudinal studies examining value and trait changes, as well as a few cohort studies that examine mean-level changes across different age cohorts in this period without being truly longitudinal. Their interesting main finding is that (a) there are changes in all traits and values that are (b) very inconsistent across studies and samples, but (c) overall these effects are very small during childhood and adolescence (Cieciuch, Davidov, & Algesheimer, 2016; Gouveia, Vione, Milfont, & Fischer, 2015; Roberts, Walton, & Viechtbauer, 2006). For example, Gouveia et al. (2015) found increases for all but one value type in their cohort study in Brazil across the age group from 12- to 18-year-olds, whereas the study by Cieciuch et al. (2016) examined both cohort and age effects among slightly younger Polish children and adolescents, and found both complex increases and decreases across a one-year interval. The meta-analysis by Roberts et al. (2006) showed increases for all trait ratings in this age range. A plausible explanation for these complex and incongruent patterns across studies is hard to find. We have to remember that we are dealing with ratings, and probably the ratings in this age period are affected by various cognitive and emotional maturation processes in addition to trait and value content. There might be considerable fluctuations during this period before some stability is reached. What is informative in this respect is the finding by Cieciuch and colleagues that value means stabilized with increasing age.

One part of the environment that appears quite important for understanding how these values might play out in modern societies is the school environment. In many countries, children and adolescents spend a significant amount of their time at school. Next to the family environment, this is one of the most important environments affecting what values might be seen as important and what traits are reinforced or punished (Berson & Oreg, 2016). Teachers set the climate within classrooms (which is also affected by the larger socioeconomic environment within which schools are situated). Teachers reward desired behaviors and punish behaviors that are undesirable. Through their actions and how they interact with students and other staff, they also set examples and act as role models who communicate values that are seen as important (Schein, 1985). In terms of the overall school context, principals are particularly crucial because they set their schools' overall strategies, recruit and select teachers, and influence the learning environment. Indeed, the importance of principals was demonstrated in a large study with nearly 50,000 schoolchildren across 252 schools in Israel (Berson & Oreg, 2016). Over a 2-year period, the values that children rated as important became more similar to the values that the school principals endorsed. These effects were also mediated by ratings of school climate (provided by teachers). These indirect effects via school climate were significant for all higher-order values except for conservative values. However, to put the size of these changes into perspective, the effects were relatively small, explaining a maximum of 10% in the value ratings of children. This is probably not surprising, given the relatively modest (and probably unstable) fluctuations that occur at this stage.

YOUNG ADULTS: FROM SCHOOL TO JOBS AND UNIVERSITY

These small size of the changes during childhood and adolescence is even more striking when compared with the more profound and differentiated value and trait changes that start to emerge in early adulthood. This is the period when people begin to move out of their childhood homes and start their own families, finish mandatory education, and start out on their own career and life paths. To understand the emergence of personality, therefore, it is important to look at later stages than teens and adolescents and at those periods when people take on new social roles. In this period from about age 18 onward to the late twenties, much larger changes in value and trait ratings appear (Cieciuch et al., 2016; Gouveia et al., 2015; Roberts et al., 2006).

Normative Transitions and Social Investment Theory

These changes can be seen as quasi-normative in that most people will exhibit similar patterns of maturation during a particular life stage. To the extent that these transitions occur for all individuals within an age group roughly at a similar time point (e.g., mandatory schooling at a certain age), these transitions are normative in character.

Normative changes can be driven by both genetic changes and changes in roles and life tasks that are typical at different life stages. McCrae and colleagues (2000) have been the strongest proponents of the importance of biological factors for driving changes. They argued that both adolescence and early adulthood are crucial periods in which these changes are taking place. Information encoded in our genes are thought to drive the changes that can be observed during this important developmental period. In contrast to this biological view, social investment theory (Roberts, Wood, & Smith, 2005) proposes that changing social roles drive personality changes because of different expectations, skills, and life tasks that individuals face. These role changes typically come later than adolescence. New roles come with different types of expectations, and the individual will be rewarded for meeting them or punished for not doing so. These rewards may not always be material, but could involve higher social status, esteem, and reputation. Similarly, punishment might be social ostracism or social isolation. Hence, role-congruent behavioral traits and values are likely to be positively reinforced, and role-incongruent behaviors are likely to be negatively reinforced (Lodi-Smith & Roberts, 2007).

So far in this book, the social environment has taken a secondary role. Part of the reason is that variables related to systematic variation in social environments are difficult to define and measure across different ecological environments. Indeed, the various processes that I am trying to illuminate here (e.g., differences in values and traits, which are often operationalized as individualism versus collectivism values) are often taken as predictors to explain the very same phenomenon (e.g., differences in personal values and traits). When drawing on such studies, it is easy to engage in circular arguments. Nevertheless, the social environment is an important environmental context that needs to be examined. To avoid circularity, here I focus on studies that have examined social context in other ways than through value or trait ratings, primarily by examining social roles that people assume as they enter new environments (e.g., becoming a university student, entering work, being promoted to a new work position) or significant life events that affect the life of a person both positively (e.g., starting a new relationship, getting married, becoming a parent) or negatively (e.g., a death of a close friend or family member, losing a job, incurring an illness or injury, getting divorced, or breaking up a relationship).

These two major explanations – changes in biological processes and changes in social roles – assume a normative character of change during young adulthood, which should apply to all or most individuals. Yet, one of the crucial questions is whether these changes apply to all individuals within a population. It is highly plausible that at least some of these events (e.g., illness, untimely death of a friend or family member, marriage, divorce, promotion to a new position) are not normative in character in all societies. Consequently, they may affect some individuals but not others. Normative processes might also play out somewhat differently across individuals. Even within normative processes, the specific expectations within roles might differ. For example, different careers (e.g., a kindergarten teacher vs. artist or musician) require different role responsibilities and lead to different expectations about acceptable and required behavior. This obviously will lead to some diversity in developmental patterns within a trajectory that is normative for this age group.

Lifespan Development Theory

Lifespan development theory (Baltes, 1987; Baltes, Reese, & Lipsitt, 1980; Baltes & Schaie, 2013; Baltes, Staudinger, & Lindenberger, 1999) is the most frequently cited developmental theory to describe these general mechanisms of stability and change as individuals encounter new challenges and opportunities in their environment. It holds that the primary driver for personality development is adaptation to (a) the new requirements of a social role and (b) the general situation within which an individual is operating. For example, vocational training is characterized by greater rigidity and expectations of obedience, classes are mandatory and highly structured, and much learning takes place on the job with tasks that are hands-on; in contrast, university study is more flexible, classes are not mandatory, and learning activities are typically more unstructured and abstract.

Lifespan development theory also makes some interesting predictions about the relative stability of personality and the rate of change in personality. Early adulthood is considered to be a dynamic stage as individuals move into new roles across a number of different life domains (family, work, social life) that require new adaptations to the requirements and expectations associated with each of the roles. However, the rate of change is going to diminish as people grow in these roles. In addition, although individuals may experiment for a while with different living and working arrangements, over time each individual is likely to select a particular niche that is most suited and optimized to his or her needs and abilities. Major changes later in life are thought to occur mainly in relation to the need to compensate for diminishing abilities as the body ages. However, this is jumping ahead because we are still concerned with the changes happening during early adulthood.

Normative Changes in Values and Traits during Early Adulthood

Current data provide support for both a normative change perspective affecting all individuals within this age range to a similar degree and for idiosyncratic changes that occur with individual transitions into particular roles and socialization into particular environments with certain expectations and norms. Focusing on the normative changes, the aforementioned meta-analysis of personality changes by Roberts et al. (2005) reported an increase in instrumental trust orientations in particular, marked by increases in assertiveness (which captures social dominance) aspects of Extraversion (see Table 3.2) and significant increases in Conscientiousness. There were also overall greater shifts toward behavioral approach motivation during this period, as individuals rated themselves as increasingly higher in both Openness traits as well as in various aspects of Extraversion. Cohort studies focusing on values found similar trends in this age group (Gouveia et al., 2015; Milfont, Milojev, & Sibley, 2016). Individuals in this age group generally report comparatively higher Openness to change and Self-Enhancement-type values. Michele Vecchione and colleagues (Vecchione, Schwartz, et al., 2016) reported on a three-way study of Italian young adults that reported value changes across an 8-year period. They found linear increases in Benevolence values over this period, but significant decreases in Achievement values. Vecchione and colleagues noted that the decrease in Achievement values was unexpected, but might be partially related to the dire economic situation in Italy, which may blunt any high vocational aspirations of young adults. At the same time, conservative values and Power values significantly increased, with the largest increase occurring early on in this period. No change was found for Openness to change values.

The overall pattern in this age group suggests that instrumental control cooperation becomes important. Power values and Extraversion (especially assertiveness and the social-dominance-related aspect) become more central. The changes along the behavioral approach and avoidance motivation dimension are more complex. The economic opportunities available to young adults may determine whether traits and values toward the behavioral approach (Openness to change values, Openness traits) or behavioral avoidance motivation end (conservative values, Conscientiousness) are more adaptive and therefore salient.

Cultural Norms, Specific Role Transitions, and Life Events

The discussion so far neglected the larger cultural environment, but the study by Vecchione, Schwartz, et al. (2016) foreshadowed that the local context may make a difference. More broadly, what is normative in a particular context might vary quite a bit from society to society. Wiebke Bleidorn and colleagues (2013) examined differential change effects in a sample of 884,328 individuals from 62 countries who had responded to online personality surveys. To examine country-specific age norms for starting a family and a career, they drew on secondary data such as the average age of first marriage and the number of years of compulsory education. Overall, the authors found strong evidence of general normative maturity patterns (in line with normative cohort effects that are independent of culture). Yet, some of the effects varied by country depending on the country-specific age norms of when a person should get married and have kids and when to finish education and start to work. Younger age norms for family and work were associated with earlier reductions in Openness and earlier rises in Conscientiousness values. This suggests that the earlier individuals are settled into their career and have their own families, the earlier they show these societally normative changes in personality traits. The important point here is that these age norms differ across societies and cultures.

At the same time, there is also much variability around these normative trends, supporting the proposition that individuals do not necessarily follow the general pattern found in the population. There are clusters of individuals who show different trends over time that match their life paths. For example, a 4-year longitudinal study of students making the transition either to university or vocational schools (Lüdtke, Roberts, Trautwein, & Nagy, 2011) showed that students choosing a vocational career path reported greater increases in Conscientiousness and smaller increases in Agreeableness over time compared to students who went to university. Another study following a small cohort of women from career entry to midlife (Roberts, 1997) demonstrated that one's work environment has significant effects on personality development. Women who spent more time working (measured as the number of years working full-time) had increased self-rated agency (part of Extraversion). Women who advanced in their career and moved up in the social hierarchy at work also become more conscientious and reported higher Extraversion over time. Corroborating these general patterns, greater involvement of individuals in any important social roles within their respective community or at work also is associated with changes in behavioral traits (Lodi-Smith & Roberts, 2007). Thus, rather than being fixed, traits and values do change with the requirements that accompany particularly roles into which individuals move. There are undoubtedly selection processes (e.g., individuals choose certain roles that suit their personality; see the later discussion), but once an individual assumes a role, longitudinal studies clearly demonstrate that further socialization takes place that shapes what a person values and how he or she behaves across situations (their traits).

In addition to these shifts associated with role transitions, individual life events also affect how people develop. The earlier cited study that followed German students as they moved to university or vocational schools also included a list of significant life events. The authors found small but significant changes in self-reported traits over the 4-year time period in response to those life events. For example, positive life events (e.g., starting a new job, becoming involved in a romantic relationship) were associated with greater Agreeableness but also more Conscientiousness at the next measurement point. Similar effects have been reported in other studies in relation to people experiencing depression, which later led to changes in Neuroticism levels (Rosenström et al., 2015). Even events such as going abroad to study for a brief period can have effects on personality. Studying abroad requires greater social skills and actively approaching novel situations. Not surprisingly, a number of studies have found that students going abroad reported higher Extraversion, Openness, and Agreeableness, as well as lower Neuroticism, on their return, although not all these changes occurred consistently in all studies (Niehoff, Petersdotter, & Freund, 2017; Rosenström et al., 2015).

These studies suggest that individuals change in response to the requirements imposed on them in their environment. Yet, individuals also have agency and can influence their environment at least to some extent. For example, young adults in Western societies have some autonomy over their circle of friends and can choose with whom to interact and how often. A large German study using event sampling (individuals reported what they were doing right then and what they were planning to do in the next hour) supports these speculations. For example, young adults who reported high levels of Conscientiousness were more likely to work in the next hour, whereas young adults reporting high Agreeableness were more likely to plan being with their family. The aforementioned study by Niehoff and colleagues (2017) found that students higher in Agreeableness and Openness traits were more likely to go abroad to study. This suggests a reinforcing cycle between traits/values and the environment, which might be particular important during young adulthood when individuals explore possible roles and responsibilities within their local environment.

These findings suggest a broader contingent nature of personality change in young adulthood (Roberts & Davis, 2016). Personality changes in young adulthood are not necessarily only driven by external demands and concrete experiences, but may also be shaped by the perceptions of the individual of how well she is operating in an environment and therefore is selecting environments and paths that are deemed as important and relevant for her. An individual who is interested in meeting new people and is curious to explore the world (high Agreeableness, Openness/Openness to change values) is likely to choose studying abroad. To the extent that the student is now perceiving that she is handling the stay abroad well (with all its new experiences and social interactions), her self-perceptions of her traits and values are calibrated to incorporate these experiences. Therefore, it is the subjective experience of roles and events that might drive the calibration and fine-tuning of personality and value during young adulthood. Roberts and Davis (2016) call this the "facultative calibration model of personality development in young adulthood."

In summary, transitions into new roles during early adulthood are associated with the most profound changes in traits and values. The most likely explanation in line with lifespan theory is that these new roles come with expectations and norms about how to behave and what to value, which through repeated positive and negative reinforcement will lead to adaptations and consequent changes in the personality of an individual. Focusing on the general trends in the underlying dynamics, the overall evidence suggests that there are profound shifts toward more instrumental cooperation orientations in this time period. There are also initial shifts toward more behavioral approach motivation, which later seem to change depending on both the career progression and family situation of the individual.

PERSONALITY CHANGES LATER IN LIFE

One of the dominant views for a long time was that values and traits are set in plaster in adulthood and will not change much more (McCrae et al., 2000). Recent research has called this into question. Although studies do show that most changes occur during childhood, adolescence, and early adulthood, there are likely to be further changes in traits and values, in particular if individuals change roles or encounter new environments later in life. These changes are typically not as dramatic and large as in the earlier parts of a person's life, but they are not to be ignored.

Two particular changes are especially important to mention. The first are degenerative processes associated with aging. As people age, they are more likely to become sick, their cognitive processes change and slow down, and their body is not as strong and agile as before. This means that individuals need to start compensating for these aging processes (Baltes & Lindenberger, 1997). In general, these processes can be best managed by reverting to a more cautious, behavioral avoidance approach. Indeed longitudinal studies have found that over time individuals become more cautious, less outgoing, and less adventurous (e.g., Gouveia et al., 2015; Milfont et al., 2016; Roberts et al., 2006).

Second, as individuals age their social role within families and the larger social network changes. Children leave the house, and responsibilities shift toward taking care of grandchildren and providing advice to younger community members. These changes might be associated with a move toward more trust-oriented cooperation. Older adults often enjoy higher social status, which may require less monitoring of social interactions and therefore greater trust that others will treat one fairly. At the same time, the greater social responsibilities in taking care of grandchildren and taking on social community roles (e.g., as wise elders, volunteering for social causes) also require more trust-based cooperation. In general, these shifts are borne out in longitudinal and cohort studies (e.g., Gouveia et al., 2015; Roberts et al., 2006).

In addition to these general changes that are normative in character, individual life events are again likely to play a significant role. If presented with major loss or experiences of depression or anxiety, individuals in Western societies often seek therapy. A recent meta-analysis of intervention and therapy studies found that personality changes in response to therapy (Roberts et al., 2017). The largest changes were observed for Neuroticism and Extraversion, but smaller, less consistent changes also occurred in other personality traits. The most interesting finding was that these clinical interventions resulted in quite rapid change: people undergoing therapy reported higher Emotional Stability and Extraversion in less than 8 weeks. This is quite a profound change, which seemed to remain relatively stable over time after finishing therapy. One take-home message from these clinical studies is that relatively small overall changes in a population may be accompanied (or even driven) by more rapid changes in smaller subsets of the population. A myriad of implicit and explicit processes are likely involved in personality and value change of adults as they react and interact with their local environment (Wrzus & Roberts, 2016).

SUMMARY OF TRAIT AND VALUE CHANGE

The evidence presented here shows that both values and traits change over the lifespan. Early infancy and childhood appear crucial for setting the overall trajectory of an individual's development. At the same time, there is still large plasticity throughout a person's life. The most profound changes in both values and traits occur during early adulthood, when individuals make the transition into new roles and responsibilities. Yet, personality changes can occur throughout a person's life, often in response to salient events that force an individual to reflect on his life or via changes in roles and social expectations. The environment plays a crucial role because what roles and responsibilities are salient will differ in multiple ways across different economic, social, and historical conditions.

To explain all these effects using a metaphor, we could return to the car example that I introduced in Chapter 1. Individuals have different general predispositions. The best equivalent of this would be different types of cars – there are racing cars, SUVs, and family cars. These cars need to built and assembled (think infancy). How much care was taken when assembling them will definitely determine how well they run later. The available resources and tools also might influence what type of car leaves the assembly line. Imagine a company that, because of scarcity, replaces high-quality components with lower-grade material – the car will still function, but might look and operate differently. This resembles somewhat the effects of economic conditions on the very structure of values and traits (see Chapter 7). Once these cars are at the car dealers, probably not much is happening. This is the childhood and adolescence of a car: only relatively minor systematic changes seem to be happening during this period. However, the real tests will come when cars are sold, and they become operational (cars take on their first proper roles). This is the early adulthood of a car. The general environment and the tasks that they have to do will determine how well and how long they will run. A family car that is driven around rugged unpaved mountain roads in the rural countryside is likely to go slower and break down more easily than a family car that is used in the city. A racing car will be able to go much faster, but road conditions and traffic laws will constrain its speed. A racing car has a higher potential to go fast, but it may even drive slower than a family car or SUV when driven on cobbled streets or on a bumpy rural road. Therefore, the environmental conditions and specific tasks will strongly affect the performance of any car. I could take the metaphor further (think of garages as therapies in case of breakdown, car tune-ups as acquiring new competencies or calibrating traits and values, etc.). Nevertheless, I hope some general ideas of personality development become clearer when transferring these ideas to a cultural invention that has become so central in our modern lives. In the beginning of this book, I suggested that we should look past the exterior and examine the engine under the hood. In addition to examining the engine, we also have to look at how carefully it was manufactured and maintained, as well as the road conditions (the life experiences in different social and economic contexts), to understand how it might perform in any particular situation.

HOW DO WE KNOW WHAT ARE OUR VALUES AND TRAITS?

So far, I have focused on self-ratings of values and traits, assuming that individuals always know who they are and what they do. But this begs the bigger question of how we might be able to consciously develop an awareness of our values and traits. Nearly all the research on human personality relies on some form of reflection, introspection, awareness of one's actions, and some personal insight into what a person is like. Infants are certainly not born with a clear sense of what they think is important in their lives nor how they think in everyday situations. This sense must develop over time. One plausible account of how this sense develops has been put forward by Daryl Bem under the name of self-perception theory (Bem, 1967, 1972):

> Individuals come to "know" their own attitudes, emotions, and other internal states partially by inferring them from observations of their own overt behavior and/or the circumstances in which this behavior occurs . . . the individual is functionally in the same position as an outside observer, an observer who must necessarily rely upon those same external cues to infer the individual's inner states.
>
> (Bem, 1972, p. 2)

Values in particular are often poorly cognitively elaborated – implying that people very rarely have rational justifications for or against a particular value (Maio, 2010; Maio & Olson, 1998). As a consequence, it is likely that people infer their values from internal observations of their behavior. Traits by definition are behavioral regularities that require an individual to monitor, remember, and integrate these various behaviors into coherent accounts of who one is as a person.

An important element of self-perception is that actions are only used as guides for updating internal mental representations of values and traits if they fall within a range of behavior that is not too discrepant and different from other behaviors that an individual has engaged in previously (a so-called range of acceptable behaviors) and the specific action cannot be attributed to external constraints or forces (Fazio, Zanna, & Cooper, 1977). Thus the source of perceived agency is a crucial ingredient for both values and traits. Put differently, individuals are likely to infer their values and traits from their voluntary actions, but only if those actions fall within a general latitude that the individual finds subjectively acceptable and they cannot be explained by external (e.g, I had to act like this because of a law) or internal (e.g., feeling ill) conditions. Being sick and lying in bed are not typical behaviors for an outgoing extrovert. Likewise, being forced to dance with a new colleague in a social function is unlikely to change the self-perception of a shy person. Self-perception theory might be an important theoretical piece in the larger puzzle of how people develop their personal values and internal representations of traits; in short, their self-awareness of their personality. How could this play out in a developmental sequence?

A BROADER NEURODEVELOPMENTAL PERSPECTIVE ON VALUES AND TRAITS

In Chapter 5, I briefly referred to some of the neuroscience approaches that tap into the underlying motivations guiding traits and values. Equipped with these insights, it might be possible to speculate how values and traits first form and crystallize during a person's early development (Brosch & Sander, 2013). Both values and traits are linked to brain reward networks (associated with making economic decisions, reward sensitivity, and evaluating the valence of a reward) and social-cognition-related networks (involved in forming impressions of others and evaluating the needs and goals of others, also called Theory of Mind). Brosch and Sanders also highlighted links to related networks involved in processes of behavioral inhibition and error tracking (which are compatible with behavioral avoidance motivation) and to areas of the brain that integrate self-related information and abstract long-term goals (Brosch, Coppin, Scherer, Schwartz, & Sander, 2011).

These neuroscience studies provide an interesting angle for evaluating the causal relationship between values and behavior. Brosch and Sanders (2013; see also Brosch et al., 2011) drew on self-perception theory in their account of value development. Subjective value hierarchies are likely constructed and adjusted when genetically or epigenetically driven behavioral predispositions are repeatedly reinforced during early life experiences. Behavior is also reinforced by others, which creates anticipations in the mind of the little toddlers that then reinforce and strengthen reward networks. To figure out how to behave, children will integrate information about their past behavior and how others around them reacted to their actions into scripts of what works when. Obviously this ability relies very much on the motivational predispositions and innate capabilities of a child – not everyone is equally good at interacting with others or feeling secure enough to trust others. In short, infants are likely to construct and adjust both value hierarchies and personal representations of traits based on their own observations of their behaviors (Bem, 1972). An infant predisposed to be outgoing might observe herself engaging with other children and adults to a greater extent than other children and, on reflection, will form a concept that she values meeting others and is relatively sociable and outgoing.

There is good evidence that stable behavioral dispositions that predict personality traits later on in life emerge even within the first few months postpartum (Caspi et al., 2003; Moffitt et al., 2011; Slobodskaya & Kozlova, 2016). Similarly, caretakers might reinforce and reward certain types of behaviors ("It is so nice that you play with all the other children"), leading to implicit evaluations associated with observed self-behaviors. These are the seeds of what will become mental representations of values and traits. The available research evidence demonstrates that even preschool children are able to differentiate between human values in line with the Schwartz value theory (Döring et al., 2015) and Big Five theory (Markey, Markey, & Tinsley, 2004; Markey, Markey, Tinsley, & Ericksen, 2002; McCrae et al., 2000).

Over time, these emerging value hierarchies and trait representations become integrated into a more and more coherent self. Once this integration occurs, mental representations may become self-rewarding even in the absence of concrete choice or reward situations. Especially for values, an adolescent may actively seek out situations associated with those values that are experienced as rewarding, further reinforcing these values. The associated behavior is simultaneously integrated into trait scripts (e.g., I am an extrovert, I like to be around other people more than others), leading to reciprocal value-behavior influences. Continuing observations of behavior and reinforcements of these behaviors by the social environment are compared with the internalized value hierarchy and trait representations and are periodically updated and revised, if necessary. Therefore, my proposal is that individual differences in the underlying biological systems (Figure 6.3) are expressed in behavior, which in turn is reinforced, and the perceived behavior and the reinforcements form the beginning of what will become mental representations of values and traits. The available evidence is very much in line with these arguments, showing that both values and behavior influence each other over time (Vecchione, Doering, et al., 2016).

Reconnecting the developmental perspectives that I outlined earlier with the structural fluctuations in different economic environments that I discussed in Chapters 7 and 8, we could speculate that the environment in early infancy and childhood is crucial for the full development of value and trait representations. As we found in our study of Brazilian physicians, the economic environment during childhood appears to be particularly important for setting the basic parameters that allow the differentiation of values (and possibly the mental representations of traits). Similarly, the social environment in which a child and adolescent is located will influence value and trait priorities, as we had seen in the large internet personality study conducted in 62 countries (Bleidorn et al., 2013).

SUMMARY

Personality is adaptive and changes with the requirements and expectations that an individual is faced with over the lifespan. Returning to Tina, she is highly involved in social roles that are important for her religion. She has to take care of her small family in an environment that is relatively economically deprived. Tina also came from a family that was relatively poor, and she had only a basic school education. As a consequence, she may rate conservative values as more important and report higher Conscientiousness and lower Openness to change values and traits. She should also be more self-enhancing and slightly more socially dominant in her value orientation and traits, because relationships are important in this context to maintain high social standing and thereby obtain resources that are important for her and her family's survival. Sandra should show similar patterns of traits and values, yet her transition into a highly competitive job was a central force in shaping her values and traits in this direction. She had to perform at high levels, which required her to become not only highly conscientious but also very achievement and dominance oriented to survive in her competitive work environment. Experiences at different time points in life may end up showing the same trait or value pattern (the concept of equifinality).

Personality change is thought to be adaptive, explaining why there is some change and plasticity in the face of social and economic conditions, even though there is also relative stability over time in the life of an individual. These changes unfold over the lifespan, but they have been shaped by longer evolutionary trajectories. In the next chapter, I engage with the larger evolutionary picture, examining both long-term evolutionary trajectories and specific interactions between genes and environment as they play out over time. I also examine the question whether the profound and quite revolutionary changes over the last 200 years (in particular in medicine, but also the technological changes that took place during the Industrial Revolution and the beginning of the information age) have resulted in an end of evolution for personality.

11

A Second Look at Evolution, Genes, and Personality

The previous chapter discussed personality development over the lifetime of an individual. In this chapter, I zoom out and consider the development of traits and values over evolutionary time. Both traits and values are heritable, with some estimates suggesting that between one-third and one-half of variability in our traits and values is caused by genes (see Chapter 5). If personality variability between individuals in both traits and values is adaptive and allows individuals to survive and pass on their genes to the next generation, then at some point the best combination of traits and values would prevail (in technical terms: become fixed) in any human population; variability in traits and values between individuals would then be eliminated, and as a consequence, there would be no meaningful differences between individuals. In other words, there would be no personality differences between individuals. If this was the case, you would not be reading this book because there are clear differences in traits and values between individuals and populations that need explaining. Why do we have variability in traits and values? What are the most likely evolutionary processes that explain why we have this diversity in traits and values? These are not trivial questions. I outline some of the major theoretical arguments and review some of the available evidence for the various accounts. Given the strong heritability of personality suggested in twin studies, most of the research to date has focused on genetics, but I also briefly touch on culture and the interplay between culture and genetics.

EVOLUTIONARY GENETICS OF PERSONALITY

When discussing the evolutionary dynamics of personality, the obvious starting point is to focus on genetics and different selection models that can account for variability in genotypes and phenotypes in a population. This is nevertheless a controversial area and different viewpoints exist. I therefore outline the major models and their implications for variability before turning to the question of which model is best supported by empirical evidence. Most researchers typically consider three major models of selection, with a number of submodels within each (see Table 11.1; Penke, Denissen, & Miller, 2007; Verweij et al., 2012).

Selective Neutrality

The first model is selective neutrality, which implies that genetic variations that give rise to personality phenotypes do not affect fitness and therefore gene variants randomly drift within a population without being affected by selection. In Chapter 1 I introduced the concept of random drift. Under genetic drift, we would expect that one variant of a polymorphism would move toward near-fixation within a population (i.e., almost 100% of the population would have the same genetic variant). This gene might affect the phenotype, but if this does not translate into any fitness advantage, it is considered selectively neutral (hence the name). However, new mutations always emerge and therefore will result in some genetic variation in a population (a so-called mutation-drift balance). Overall, in this model, differences in personality have no implications for fitness, which means personality differences have no effect on survival and reproduction rates.

However, this assumption of no fitness effects of traits or values is unlikely as we have seen in previous chapters. The most frequently discussed example showing explicit fitness effects of traits is Extraversion (Nettle, 2005, 2006). Individuals high in Extraversion are more likely to gather more resources through the exploration of the environment, as well as rise to high social rank through dominance behaviors and forming more effective social alliances. These outcomes then give extroverts a greater chance to find more mating partners and have a higher number of offspring, who in turn are more likely to survive because of the greater resources available to their extrovert parents. Therefore, high levels of Extraversion increase fitness by leading to higher chances of survival and reproduction. This relationship can be observed both in industrialized and traditional societies (Alvergne, Jokela, & Lummaa, 2010). At the same time, there are costs involved with being an extrovert. Exploring the environment carries the risk of injury or death. Imagine the environments in which our ancestors roamed as hunter and gatherers. Going out to hunt animals always carried the risk of encountering some dangerous predators (e.g., lions, tigers) or just physical injury

TABLE 11.1 *Overview of Different Genetic Models of Personality*

Model	Description	Number of Active Genes	Empirical Support
	Selective Neutrality		
Selective neutrality	No selection on phenotypes, different variants of genes provide equal fitness benefits	Unknown	Not compatible with existing data, mainly because there are clear fitness advantages for traits/values in different environments
	Mutation-Selection Balance		
Mutation-selection balance	Balance between deleterious alleles being selected out of a gene pool and new deleterious alleles emerging through mutation	Many	Compatible with genetic data showing many genes contributing small amounts of variance to phenotypes
Condition dependence	Trait phenotypes are calibrated on the physical conditions of the body, but selection on body morphology rather than traits	Many	Theoretically compatible with existing data, but cognitive mechanisms unknown
	Balancing Selection Models		
Balancing selection	Multiple alleles maintained in the gene pool at higher frequencies than expected by genetic drift	Variable	Depends on specific variant
Environmental heterogeneity	Fitness of a trait varies across time or environments	No predictions	Plausible but currently popular models focus on single or small number of genes (e.g., 5HTT & DRD4 transporter genes); empirically unlikely
Negative frequency-dependent selection	A rarer phenotype has higher fitness	Few	Unlikely, given that traits and values are influenced by a large number of genes
Pleiotropic balancing selection	Genetic variation is side effect of pleiotropy whereby a gene influences multiple unrelated phenotypes	No predictions	Plausible

caused by bad luck (e.g., falling from a tree or cliff, being bitten by a poisonous snake, becoming involved in a hunting accident). The greater opportunity to find mating partners may also have a downside: having more options is likely to increase the instability of any romantic relationship. There is ample evidence in modern societies that supports both the fitness benefits and costs of Extraversion (Nettle, 2005).

Similar processes might also operate with Neuroticism, which is arguably one of the more difficult traits to explain from an evolutionary perspective. Neuroticism may be beneficial initially because emotional expression (e.g., increases in crying frequency if the baby is experiencing some discomfort) forces mothers to pay more attention to infants (Soltis, 2004). However, Neuroticism later in life may lead to anxiety, depression, health problems, and premature mortality (Ferguson, 2013) and therefore may reduce chances of survival and finding mating partners. Conversely, in a threatening environment, individuals high in Neuroticism might have higher chances of survival due to their cautious nature and avoidance of highly threatening situations, as well as greater vigilance when exploring environments (e.g., spotting potential dangers earlier than other individuals who are less cautious). In contrast, in more benign environments, individuals high in Neuroticism might miss out on potentially important resources, therefore putting them at a disadvantage. Focusing on the social contexts, Neuroticism might also be beneficial in that individuals with higher levels are constantly scanning the social environment for signs that they are socially excluded. Humans are a hypersocial species, and given our comparatively weak bodies, our only chance of survival in evolutionary history was to bond together (the proverbial safety in numbers). High levels of Neuroticism might be adaptive because constant monitoring of relationship quality is likely to help individuals protect limited relationship opportunities and therefore survival in a hostile environment. Returning to the ontogenetic processes discussed in the previous chapter, Neuroticism levels are likely to increase after experiencing negative life events as shown in the German longitudinal study of individuals as they transition from school to work or university (Lüdtke et al., 2011). Yet, high Neuroticism is also associated with decreased physical and emotional health and is likely to strain social relationships, increasing its costs.

With this balance of costs and benefits, it might be plausible that the net benefit of each trait is close to zero (MacDonald, 1995). However, this assumes that individuals are operating in a single environment. Costs or benefits might vary depending on the environment, causing different phenotypes to be more or less advantageous. This will be discussed later in the section on the various balancing selection models. Furthermore, if the same genes affect multiple personality traits (as could be the case given the assumed underlying major behavioral approach versus avoidance systems that modulate various different traits), then selection might not be acting on these genes, even though there are fitness benefits associated with each phenotype (and the underlying genes). Therefore, selective neutrality is a model that is plausible from a theoretical perspective. Nevertheless, given the demonstrated fitness benefits, it is not likely to fit the data well. Neutrality is typically the default assumption (no selection has happened), and researchers have to show that any genes have been selected for their fitness benefits. This is where the next two selection models come in.

Mutation-Selection Balance

Under mutation-selection balance, new mutations constantly emerge, but deviations from an optimal trait level (across all possible environments) are strongly selected against. That is because new mutations are likely to unbalance a finely tuned biological system and increase noise. Our brains are enormously complex systems that require just the right balance between millions of cells that communicate with each other for the body to survive. Any small mistake anywhere along the development of an individual caused by tiny shifts in the genetic code is potentially disastrous. If these deleterious effects are very harmful, an individual is unlikely to survive (many new mutations end in miscarriage or early deaths of infants). This means the new genetic variant caused by the mutation is immediately selected out of the gene pool of a population. If the effects are weaker or recessive (that is, two versions of the same allele need to come together to affect the phenotype), they are less likely to be immediately selected out. However, even though these versions might continue to be present in a population, they are unlikely to rise to prominence because of the disruptive effects on the phenotype; therefore, over generations such effects will disappear. As a consequence, each one of us will carry a number of rare, recessive (weak), and mildly damaging mutations that in combination shape the biological architecture that ultimately forms our personality phenotype. This so-called mutation load refers to the total number of such alleles that have a weak, but deleterious, effect on fitness.

Initial studies suggested that a relatively small number of candidate genes regulating dopamine and serotonin processes in the brain might account for phenotypic differences in personality traits (Benjamin et al., 1996; Lesch et al., 1996; Munafò, Yalcin, Willis-Owen, & Flint, 2007; Reist et al., 2007).

However, this early optimistic view was soon discounted by large studies that were unable to identify common variants that accounted for personality variation; if they did identify individual genes, the specific associations could not be replicated in other studies (Munafò & Flint, 2011). For example, a large consortium of researchers (de Moor et al., 2015; van den Berg et al., 2016) conducted meta-analyses involving more than 60,000 individuals to identify possible genes that could account for variation in Neuroticism and Extraversion scores. Only one polymorphism in a gene was significant for Neuroticism across the whole genome, and no individual polymorphism was significant for Extraversion. This seems to suggest that the traits' high heritability is due to the involvement of a large number of genes. There might be a countable number of specific genes that affect personality phenotypes, but the contribution of each gene to the overall phenotype will tend to be small (Turkheimer et al., 2011). However, given the large number of genetic loci involved in personality phenotypes, the "mutational target size" is large, and many new mutations are likely to arise across the genetic loci. This means that personality phenotypes can be easily influenced by mutations across a large number of genes, which explains the variability of personality.

One variant of mutation-selection balance is so-called condition-dependent selection (Lewis, 2015). Condition-dependent selection assumes that there is a species-wide phenotype that is fitness enhancing; for example, a larger and stronger individual can fight off predators and competitors or an individual with greater physical attractiveness is more likely to attract mating partners. These characteristics are then taken as input cues for adjusting personality strategies that would have the best cost-benefit ratio for the individual, given his or her condition. This process assumes that there are fitness tradeoffs for the non-personality-related phenotype that were present in the environment of evolutionary adaptiveness when humans developed. For example, stronger individuals are probably more likely to benefit from a personality phenotype that is extroverted, approach oriented, and sociable instead of an introverted personality phenotype. At the other extreme, an individual who is less likely to succeed in dominance contests would be better off avoiding such situations, developing a more introverted and shy personality trait profile.

This variation of mutation-selection rests on studies showing that physical body morphology is often correlated with personality traits (Fink et al., 2012; Fink, Neave, Manning, & Grammer, 2005; Lewis, 2015; Lukaszewski & Roney, 2011; Pound, Penton-Voak, & Brown, 2007). Therefore, personality phenotypes are calibrated on the physical conditions of the body, but selection is working on body morphology, rather than personality. This overall rationale is compatible with the facultative calibration model of personality development (Roberts & Davis, 2016) discussed in the previous chapter. This particular version goes beyond physical features and includes psychological characteristics such as competencies and abilities (which allow a person to master certain challenges and situations better). To the extent that such competencies and abilities are also driven by large numbers of genes, the mechanisms are compatible.

Balancing Selection

The third commonly discussed type of selection is balancing selection. Here, multiple versions of a gene (alleles) remain in high frequencies in the gene pool of a population at a level higher than would be expected from genetic drift alone (e.g., random convergence toward one version of the gene only). This is the most diverse and complex model because it assumes genetic variation is maintained in a population instead of being selected out, but the processes by which this maintenance occurs vary in form. In this section, I discuss several of these different processes, in particular those that are more likely to play a role in the development of human personality.

One likely reason for the occurrence of balancing selection in the personality domain is that fitness payoffs (survival and reproduction rates) of the personality phenotypes vary across environments. This is called environmental heterogeneity. It requires that a particular trait has no fitness advantage overall because of temporal or spatial changes in selection pressures (similar to the zero net benefit mentioned earlier). In other words, in some environments one version of the gene is beneficial, whereas in a different environment, another version of the gene is more adaptive.

A classic example from animal personality is the great tit, a common songbird in the Northern Hemisphere (Dingemanse, Both, Drent, & Tinbergen, 2004). These little birds express personality differences in how fast they explore novel environments, and this trait resembles Extraversion in humans. Birds with the fast-exploring personality are aggressive and bold in exploring new environments, but are relatively insensitive to external stimuli and tend to follow routines. In contrast, slow explorers are more passive, are relatively shy in exploration, pay more attention to external stimuli, and readily adjust their behavior to match changes in the environment. The researchers caught the birds and measured their personality styles during a laboratory task. Once released back into the wild, Dingemanse and colleagues then monitored the survival and reproduction rates of fast versus slow birds, in response to the availability of food and environmental conditions across the years. Males are dominant over females and need to protect their territory against other males. Bold and aggressive males benefited in years where food was widely available and more young birds survived, which increased the competition for territory in the following season. Therefore, aggressive males were more successful in protecting their territory against these relatively inexperienced young ones. However, in years where there was less competition, aggressive birds did not prosper because aggression often brings with it higher costs of injury and wasted energy. In those years, passive and shy males did better by avoiding conflict in conditions where there was little competition for space. For females, competition for food is important. Fast-exploring females who are more aggressive did better in years when little food was available because the costs of aggression were worth the effort in obtaining scarce food resources. In contrast, passive females did better in rich years because enough food was available, and they avoided the costs of injury and even potential mortality due to higher aggressiveness.

Moving on to humans, polymorphisms in the dopamine receptor gene (DRD4) and serotonin transporter gene (5HTT), referred to in previous chapters, are often cited as examples of how environmental heterogeneity maintains genetic variation in humans. To recap briefly, both dopamine and serotonin are neurotransmitters involved in regulating the behavioral approach versus avoidance systems in the human brain. For the serotonin transporter gene, individuals with at least one short allele of the gene are more likely to avoid risks and are particularly sensitive to negative stimuli, compared to individuals with two long forms of the allele, who are more risk seeking and tuned to positive stimuli. Chiao and Blizinsky (2010) in their seminal study argued that the greater presence of pathogens in the ancestral environment led to a fitness advantage for individuals with the short allele because they were more cautious, and as a consequence they had a greater likelihood of surviving and procreating in environments that were rife with pathogens. Therefore, the environmental pressures led to the selection of a phenotype that was more risk averse. Similarly, for the dopamine receptor gene, one particular version (called the 7R allele because it has seven repetitions of a particular base pair) is thought to be associated with more risk seeking, Extraversion, and novelty seeking. The classic argument is that this allele was favored in populations that moved to new environments in which exploration was essential for gathering resources, especially if these new environments were rich in resources and relatively low in threats. The 7R allele is more prevalent in today's European populations and may have benefited the first humans moving northward out of Africa during periods of global warming in which resources were relatively abundant. As discussed previously, these accounts are speculative because it is difficult to examine the environmental conditions occurring during the emergence of these new mutations.

A second process of balancing selection, closely related to competition, is negative frequency-dependent selection. Under this scenario, the rarer phenotype has a competitive advantage (i.e., leading to higher success rates in survival and reproduction) than the more common one. Competition over limited resources is often costly because it becomes a zero-sum game. A winner in a zero-sum game will benefit, whereas the loser typically bears all the costs (at least in terms of wasted time and effort). If costs of competition increase with the number of individuals using the same strategy, individuals using the rarer strategy may pay lower competitive costs. Therefore, with increasing competition, individuals with a different strategy may actually do better. Note that the alternative strategy is not better per se: if the frequency of that strategy increases, the costs associated with it will equally rise, turning the tables on the selective advantage of alternative competitive strategies. This classic example of this phenomenon is the "hawk-dove game (Maynard Smith & Price, 1973). Animals rarely fight vigorously, but rather use limited conflict options that are highly ritualistic and nonlethal even in situations where a lot is at stake (e.g., when defending territory, mates, or food). Such strategies are beneficial for the individual because the costs of competition are typically higher in the long run than the obtained benefits. Yet, a small number of individuals who use more lethal force might do well if the majority avoids conflicts (Dawkins, 1976). Another way to consider negative frequency-dependent selection is to think of viruses. If a virus strain is common, most hosts will develop immune reactions against the virus, making it much harder for it to survive because it will be difficult to find new hosts that have not yet developed resistance to it. In contrast, a novel virus strain will rapidly spread through a population because hosts have not yet developed immune responses to it. Applied to the personality domain, the implication is that rarer genotypes are more likely to be successful in attracting mates and reproducing. You could think of it as a "novelty wins" strategy.

Another variant of negative frequency-dependent selection is related to parasite stress. Parasite stress is a common threat that has captured the attention of biologists over many years, as discussed at various points in previous chapters. The main point for our current analysis is that parasites are often well positioned to attack hosts with common variants because they have developed specialized cellular mechanisms to attack those cells or circumvent immune responses. In this biological arms race, a host with a rare variant therefore has higher fitness because it is less likely to be infected by pathogens conditioned to attack common variants. This form of selection is called pleiotropic balancing selection. As discussed previously, pleiotropic effects occur when a single gene affects multiple different phenotypes. One plausible explanation for pleiotropic balancing selection in the personality domain is that a gene's effect on a specific personality trait is perhaps selectively neutral, but the gene is under selection in another trait in which it is expressed. This is a scenario that has been suggested in a number of recent genome-wide association studies, in which genes that strongly correlate with personality traits have been linked to immune system functioning and neural signaling (important for transmission of information in the brain; de Moor et al., 2010; Lo et al., 2016). Therefore, the selection might have happened on genes that regulate important body functions, which in turn are also associated with changes in personality-relevant processes. Pleiotropic balancing selection differs from environmental heterogeneity in that the former does imply cellular processes (e.g., parasites being particularly well designed to infect individuals with common gene variants), whereas environmental heterogeneity involves more behavioral adaptations to the environment.

EMPIRICAL EVIDENCE IN SUPPORT OF THE VARIOUS SELECTION MODELS

One promising methodology to investigate the relative plausibility of these various models for personality that has recently become available is an analysis of the correlations between genes and personality phenotypes. The different models carry with them implications of how many genes are involved, their relative frequency, and how they interact in producing a particular phenotype. For example, negative frequency-dependent selection (including some common forms of gene-culture coevolution models of personality that I discussed in Chapters 2 and 9) is only feasible if there are few genetic loci that produce a phenotype, whereas mutation-selection balance implies that many diverse genes are involved in a phenotype. Such assumptions underlying the different selection types allow us to gauge which of the models is most plausible when considering the genetic make-up of individuals in a population and the correlations between genes and phenotypes within populations. Penke and colleagues (2007) originally suggested that the most plausible evolutionary genetic mechanism for personality was balancing selection (in particular, environmental heterogeneity). They argued for this mechanism based on initial studies suggesting that only a few genes regulating dopamine- and serotonin-brain systems are important for understanding personality, together with the theoretical reasoning that fitness payoffs (survival and reproduction rates) of the personality phenotypes should vary across environments.

However, when possible evolutionary mechanisms were tested more directly in a large study involving more than 8,000 individuals, a different picture emerged (Verweij et al., 2012). Verweij and colleagues used a personality inventory that measures only Neuroticism and Extraversion (Cloninger, 1986; Cloninger, Przybeck, & Svrakic, 1991). Mutation-selection balance was the most likely evolutionary scenario to explain heritability of these two traits in their study. They found a large number of random mutations that were deleterious, but each had only small effects on phenotypes that emerged at a similar rate as they were eliminated by selection. As discussed earlier, mutations that have stronger effects are quickly purged because they will unbalance the overall biological system, and therefore an individual carrying those mutations is not likely to survive. Weak or recessive mutations with small effects are less likely to strongly affect the phenotype and therefore are not strongly selected against. At the same time, they are not likely to become common in a population because of their negative effect on survival and reproduction rates. Hence, the evidence suggested the presence of a large number of weakly deleterious genes affecting traits.

Obviously, this is a first study using a single instrument focusing on two traits only. However, it casts further doubt on some of the commonly endorsed gene-culture coevolution models for personality. Such models are only feasible if there are relatively few genes with relatively strong effects on phenotypes so that selection can work on them. Classic examples have focused on either diet or disease models, where clear fitness advantages can be established. For personality, fitness is more variable. The available genetic evidence suggests that gene-culture coevolution for personality traits and values is less plausible, at least for individual genes as commonly studied in genome-wide association studies.

Lewis (2015) maintained that evidence supports condition-dependent selection as a variant of mutation-selection balance. The major evidence hinges on the correlations between physical attributes and personality traits (and values). One current limitation with these studies is that they are based on relatively small samples (often less than 100 participants) drawn from student populations in Western societies (but see also the arguments by Roberts and Davis, 2016). It needs to be shown whether such effects can be replicated in larger, more diverse samples. At the same time, condition-dependent selection might be more suitable for supporting gene-culture coevolution models. Selection pressures changed as humans moved into new environments around the globe. Our ancestors had to survive the onslaught of diseases that were localized in their environment and had to get sufficient nutritional value from plants and animals that were available for consumption given the climate and geography. Hence, there are strong selection signals for both specific foods and immune system functioning in populations around the world. If you are from Central America, your digestive system is better adjusted to digest maize products, whereas in Europe you are more likely to digest wheat and milk products well. Such changes may then have opened the door for condition-dependent selection in the personality domain. Such changes may also be compatible with pleiotropic balancing effects. The major difference between these two models is that pleiotropic balancing effects work more at the cellular level (with traits and values being a byproduct), whereas condition-dependent selection involves a calibration of traits. At the level that psychologists typically work, the two processes are impossible to distinguish.

Once these condition-dependent selection patterns are replicated, the next step would be to explore the underlying information-processing mechanisms that help an individual calibrate personality based on physical cues and developmental trajectories. Is there a sensitive period for these calibration effects, or do they remain plastic (e.g., if an individual suffers an accident that affects the relevant cues for a personality phenotype, is the personality phenotype recalibrated?). The approaches used in the developmental literature reviewed in the previous chapter are highly relevant to tackle this question. More importantly, from an evolutionary perspective, did the cultural evolution that occurred with the transition from hunter-gatherer environments to agricultural and industrial production affect such calibration effects, and if so, then how? This is a crucial step for understanding our evolutionary past. Condition-dependent selection is certainly a fascinating theoretical proposal that requires further investigation.

BEYOND SIMPLE GENE EFFECTS: INDIRECT GENE EFFECTS, EVOLUTIONARY TIME SCALES, CLIMATE CHANGE, GUTS, AND DIET

Humans are a hypersocial species with the unique trait of possessing cumulative culture, in which cultural artifacts and traits become more complex with each new invention. These features are likely to have far-reaching implications for understanding evolutionary dynamics. For example, indirect genetic effects might be very common, yet are currently underestimated. Indirect genetic effects occur when the genotype of one individual affects the phenotype of another individual (Bijma, 2014). A classic example is the mother-child relationship: the genetic make-up of the mother may affect the fitness of a child, regardless of his or her genetic make-up. For example, having a mother with genes that make her more caring will increase the probability that the child will thrive (regardless of whether the child herself inherited those particular genes), thereby increasing the fitness of the child. In human society, these effects might scale even further. For example, I discussed earlier the study of school principals showing that the values of the school principal were correlated with pupil values two years later (Berson & Oreg, 2016). What if these effects were actually partially driven by indirect genetic effects instead of socialization, as assumed by the authors? Biologists often suggest that such effects are widespread, yet it may be difficult to prove or disprove them in the case of humans. To answer these questions, we need more interdisciplinary thinking and collaboration. Indirect genetic effects are likely to be common, but will be difficult to pinpoint accurately.

What makes these studies even more tricky is that classic evolutionary psychology assumes an ancient environment of evolutionary adaptation (Buss, 1991, 2001, 2009; Cosmides & Tooby, 1989; Tooby & Cosmides, 1995). The common argument is that psychological characteristics were set at a time when humans split from other hominid ancestors and roamed in small groups across the African savannah. We do not quite know what that environment might have been like for humans. In addition, there is plenty of evidence that biological evolution is working both slowly *and* quickly. A single event happening within a split second could radically shift the genetic make-up of a population. These events do not need to be systematic: a random event such as a natural disaster may randomly kill a substantial proportion of a sample (which is random drift or neutrality). Yet these events can also be systematic. Diseases may spread through a population, sparing only those individuals with immune responses to these illnesses. A classic example in Europe is the plague, which ravaged the population in repeated waves in the Middle Ages. There are clear selection signals in the human genome that show selection for individuals who were resistant to the plague (Laayouni et al., 2014). Some of the genes that might have been selected in response to parasites have also been implicated in mood disorders (Ferreira et al., 2008), which form the clinical end of personality traits (e.g., Neuroticism, Extraversion; Lo et al., 2016).

Our social environment also is likely to trigger selection. Chen and colleagues (Chen, Moyzis, Lei, Chen, & Dong, 2016; Zhu, Chen, Moyzis, Dong, & Lin, 2015) demonstrated that schooling has led to selective pressures and selection toward phenotypes that are more likely to do well in modern societies. Going a bit farther back in time, if we think about wars and conflicts in human society and link them to what evolutionary biologists have taught us about competition, we might speculate that extended conflicts may actually have worked as selection pressures. Events such the devastating Thirty Years' War in Europe (from 1618 to 1648) that decimated population numbers may well have worked as selective events. We can only speculate who might have been more likely to survive and reproduce – the outgoing and bold person or the shy and reserved one? This is certainly an interesting avenue for further research. Some authors have argued that wars in particular have served to self-domesticate us, making us more docile and less aggressive over the last few millennia (Turchin, 2015; Turchin, Currie, Turner, & Gavrilets, 2013).

This immediately brings the issue of culture and the interaction between genes and culture back into focus. In previous chapters, I discussed some of the classic gene-culture coevolution models that have been proposed to account for both trait and value differences across populations. These are highly exciting, yet speculative models given the available evidence, as we just saw. Further genetic work might be able to bring more insights. One avenue of research is more fine-grained examination of disequilibria in the gene distribution that can signal locations where recent selection has been taking place (Chen et al., 2016). A second avenue that might bring interesting insights is the analysis of ancient DNA (Leonardi et al., 2016; Llamas, Willerslev, & Orlando, 2017). Once skeletal samples are available from different regions and time periods from which high-quality DNA can be extracted, it might be possible to correlate polymorphisms of interest for personality research with varying environmental and social conditions over time and world regions. Especially for those genetic variants where some hypotheses about coevolutionary processes have been proposed (e.g., dopamine and serotonin regulation), such data would be most welcome to help us gain a better understanding of how we became us.

It is also important to examine the selection pressures in modern environments. Fertility rates in modern societies are changing. In Western societies, many segments of the population (especially those with higher education) are having fewer children, whereas fertility rates remain stable or are increasing in other parts of the same society and other parts of the world. What are the effects on population-level dynamics at the genetic level? Do these effects actually matter, or are social and economic effects more important? Social and personality psychologists tend to exclusively focus on socialization effects, but some of the effects might be related to changes in population genetic characteristics. Yet, this requires careful measurement of the phenotypes (requiring highly accurate measures), the social context, and the specific selection pressures (Courtiol, Tropf, & Mills, 2016).

Our modern world is also facing other challenges, most notably dramatic climate changes, as well as changes in diet and behavior. Earlier, I showed how climatic demands influence traits and values. Furthermore, personality might be linked via pleiotropic effects to genes that regulate temperature homeostasis, immune system functioning, and neural communication efficiency. All these processes are likely to be affected by global warming, which could lead to selective pressures that then indirectly lead to shifts in personality phenotypes. Independent of how these effects emerge (e.g., through physiological, cognitive, or some other mechanisms), climate change is likely to affect traits and values in the medium and long term (Van de Vliert, 2013a).

One other mechanism that I have left out of the picture so far is diet and gut microbes. There is emerging evidence that both diet and its interaction with gut microbes have significant effects on personality dynamics (Dash, Clarke, Berk, & Jacka, 2015; Foster & Neufeld, 2013; Schnorr & Bachner, 2016). The food that we eat may also affect basic brain functioning through the nutrients that it contains. For example, fruit and vegetables are rich in chemicals that are transformed into vital neurotransmitters such as dopamine and serotonin, which may then influence our traits and values. In a recent large-scale experimental study in Australia, researchers found that changing to a more fruit- and vegetable-rich diet was associated with increases in Emotional Stability-related states (Mujcic & Oswald, 2016). What we eat may have a direct influence on neurotransmitter functions, which then could scale up to changes in traits and values.

I have been concerned with how cultural differences in personality traits and values can be explained through evolutionary principles. Previous research has focused on genetic effects; more recently, these genetic effects have also been examined in light of cultural coevolution. Yet, we all eat different types of food, which have been shown to affect both our genetic make-up and the microbes in our bodies. How ironic that research has largely ignored one of the most central cultural elements that has differentiated human populations until very recently, which in turn may have a profound effect on the very core topics of this book. I decided not to delve into these issues of diet because of their speculative nature at this moment.

Yet, these are important areas that warrant further scrutiny and investigation and might well yield fascinating new insights into human psychological functioning across the globe. Maybe in the future we may be able to see evidence for the proverbial "we are what we eat".

SUMMARY

We have moved full circle in this chapter, linking the themes in this book through the long-term evolutionary lens by examining how evolutionary changes may have shaped traits and values. Lacking time machines that allow us to go back in time to examine these processes directly, the best option available to us is the evidence of evolutionary change as imprinted in the genetic record. The questions of why we have differences in personality traits in the first place and how evolution may have shaped traits (and values) have divided researchers. I focused on some of the major theoretical models proposed and discussed in the broader personality literature. These models were developed in biology, and I tried to explain their major differences without providing any of the complex equations or describing all the specifics that are necessary to fully understand these models. Therefore, biologists may feel that my descriptions are too simplistic, but I hope that these simplifications and examples helped explain the major issues and demonstrate the intricate complexities that are involved in answering these questions.

In combination with the evidence of the relative influence of genes and culture on traits and values, the overall picture points to the presence of hundreds and thousands of typically mildly deleterious mutations in our genes, each contributing a tiny amount of variance to our personality traits and values. These gene effects then influence the overall fitness condition of our bodies. Depending on the environment within which an individual happens to live, these conditions may then shape behavioral responses. Our traits and values are likely to be calibrated and conditioned on what our body will allow us to do in order to satisfy our biological and social needs. A weak and feeble body may not benefit from an outgoing and dominant trait or value profile, but a shy, cautious, and avoidant orientation may allow this body to survive well enough. The diverse environments that we humans have encountered and created as we moved around the world are also likely to have played a significant role in shaping our genes and thereby our traits and values. Yet, we should not expect to find that single genes alone or in interaction with single environmental conditions shaped these complex traits or values in the form discussed in this book. Given the complexity of human traits and values, we should not look for simple evolutionary solutions in the gene pool. To identify and describe possible mechanisms of how such interactive processes may have played out over evolutionary time scales in the different environments that humans have populated, we need more interdisciplinary and imaginative research. This is one of those beautiful challenges for future research.

The patterns that I reviewed in this chapter also help add additional details to the broad-brushed picture that I painted of Tina and Sandra. Both of them will harbor many thousands of mildly harmful mutations that influence the fine balance of the interactions in their brains. Tina faced some milder health problems in her earlier life, which motivated her to seek out a traditional healer. She did not have enough money to pay for a full service. This healer did not mind and provided her with great counsel, but also persuaded her to become a spirit medium to help her to get better and pay back her dues to the gods. Her health condition, in conjunction with her financial situation, therefore set into motion a socialization process that deeply influenced her and shaped her into becoming a person who is taking a central role in her tightly knit religious community. Similarly, the dietary changes that went with this religious transformation (e.g., restrictions of meat and milk products and consuming more fruit and vegetables) also played a role. We could speculate that as an East Asian she was intolerant to milk, but she nevertheless had drunk milk because it was seen as a prestigious food item. Taking on this new role now restricted the intake of food that she was mildly intolerant of, which improved her health condition. Eating more fruit and vegetables may have changed her neurotransmitter system just so slightly, which contributed to tiny additional changes in her trait and values. Hence, the complex interplay of genes, the social and economic environment, and role transitions (including associated dietary choices) contributed to the traits and values that we encountered in the book's opening pages. Evolutionary processes at the genetic end played a role, but so did economic and social conditions. Tina is just the last person in a long line of individuals who over the last 10,000 years experienced similar problems. The aggregation of tiny effects at the genetic level in interaction with all these other variables ultimately led to the emergence of what we today see as traits and values.

12

Why Should We Care about Personality, Culture, and Evolution?

Personality is a fascinating topic, and many people are both fascinated with their own personalities and those of other people. Most of us also know one or two friends or acquaintances who love to talk and gossip about others and what they are like. This is all nice and interesting, but why should we care about how personality dynamics unfold?

I put forward a new theoretical account of personality, linking different strands of research on traits and values by highlighting potentially important underlying motivational systems. More importantly, I tried to argue that environmental resources and constraints interact in complex ways with these biological predispositions, and this interaction has far-reaching consequences for both the structural organization of personality and what traits and values are more or less dominant in one context compared to another. Our joint baggage of genes and environment shapes us and may make us distinct from each other in ways that have escaped the direct attention of researchers to date.

These processes are not just fascinating but also have considerable implications in a range of different domains of life, from work to health to economic development and politics. At the same time, the theoretical approach that I used in outlining these linkages may be fundamentally involved in similar processes that influence a host of other psychological variables and processes that are of interest to researchers. Personality is arguably one of the most central aspects of a person and is likely to shape how she or he will think, feel, and act in a large number of domains and spheres of life. In this chapter, I highlight some of the possibilities and implications of the model outlined in this book.

PERSONALITY AND BEYOND: HOW CIRCULAR STRUCTURES MAY UNIFY PSYCHOLOGY

My main starting argument was that values and traits in the Big Five tradition are more similar than most people would assume and are systematically linked through some shared underlying biological processes. As a starting point, I argued that behavioral approach versus avoidance systems and an instrumental control versus trust-based cooperation system are likely candidates for organizing the interrelationships between traits and values. Other systems might be involved, but these major systems are a good starting point to help us understand how personality dynamics unfold. These underlying motivational systems are likely to influence a host of other attitudes and belief systems of us humans. Diana Boer and I made a start in mapping out some possible links using a circular structure (Boer & Fischer, 2013). We were interested in examining whether such a structure could help us understand how social attitudes are organized. If these motivational systems are really that central for values and traits, then they are also likely to relate systematically to the thoughts and attitudes that people have about all sorts of social issues. We examined all available studies that had reported correlations between the 10 value types and social attitudes. We then grouped all the different social attitudes to the extent that they might tap into one of the major value orientations in Schwartz's (1992) value model. The results were overall consistent and coherent (see Table 12.1).

Let me break down some of these findings. Not surprisingly, positive attitudes toward cooperation and rejection of violence were strongly and systematically associated with the underlying cooperation axis that organizes values. If people are oriented toward more trust-based interactions, they also evaluate cooperative and altruistic acts more positively and acts of violence and antisocial behavior negatively. More interestingly, many of the social attitudes that currently polarize the industrialized world are also systematically linked to the two major dimensions. Individuals who are more behavioral avoidance oriented (they endorse conservative values) are more likely to strongly hold religious attitudes and beliefs and view hierarchy and conservative political issues positively. At the same time, these individuals are more likely to reject environmental policies and environmental protection. All of these issues are at the center of many of the current political stalemates in Western societies (and also many developing countries). These debates tap into core aspects of what people are like, what motivates them, and how they act, feel, and think. These deep divisions obviously raise

TABLE 12.1 *Overall Pattern of Results Showing the Linkages between Values and Social Attitudes*

Prediction	Result
Prosocial attitudes, including attitudes toward cooperation and altruism, as well as toward violence (reverse), are positively associated with self-transcendence values (cooperation axis).	Supported
Environmental attitudes (including attitudes toward sustainability and pro-environmental activities) are positively related to self-transcendence values and openness values (cooperation axis, behavioral approach motivation).	Supported
Religious attitudes and beliefs are positively associated with conservative values (behavioral avoidance motivation).	Supported
Attitudes toward national identity and patriotism are positively associated with conservative values (behavioral avoidance motivation).	Not supported
Attitudes toward authority and political attitudes are positively related to conservative values and self-enhancement values (behavioral avoidance motivation, cooperation axis).	Supported

fundamental questions of how they can be bridged so people can come together and negotiate workable agreements. Can you come to a political solution if people at opposite ends of the debate see the world quite differently – either as an open world to be explored, discovered, and enjoyed or as a world full of dangers and threats that needs to be tightly controlled and regulated?

I believe by understanding better the underlying dynamics of personality, we may arrive at options for solving some of these divides. Next I highlight some of the theoretical implications of the integrated model presented in this book.

Much of psychological research has focused on isolated constructs, variables, attitudes, beliefs, traits, values, and behaviors without considering how they may be linked or related to other psychological processes. Some of these barriers to thinking holistically are created by the tools that researchers are familiar with, such as factor analysis. Data analytical reduction techniques are great statistical tools for reducing the complexity of data, but they are not well suited for identifying underlying interrelations and dependencies. Even if a researcher assumes correlated factors, they are typically extracted and considered in isolation from all the other variables that also form part of the input data. The recent work on higher-order personality factors discussed in Chapters 2, 3, and 4 is an important step away from these practices. Other approaches exist that can help us better understand complex interdependencies. Multidimensional techniques similar to those that I used throughout this book hold much promise in providing a more comprehensive picture of psychological dynamics.

My work on attitudes with Diana Boer (Boer & Fischer, 2013) is just one example of what might be gained by taking a more holistic approach. Various attitudes that previously were treated as independent now seem to be related via a limited set of underlying motivational systems. For example, we can predict what attitudes are likely to vary together and how persuasion efforts targeting one specific attitude (e.g., climate change) might also affect another attitude (e.g., prosocial attitudes). Equally, a systems view helps us identify central or core attitudes that more directly link into the underlying motivations that are pivotal for attitude change. At the same time, such a systems view may also help us to understand why certain clusters of values, traits, attitudes, and beliefs covary in a particular environment or among a certain group of people. These clusters might be functional for individuals or groups within their specific environment. Attempting to change such clusters of values, traits, attitudes, or beliefs (e.g., think of helping people embrace cultural diversity or become more environmentally conscious) might fail if there is no appreciation of the developmental and environmental factors that led to the emergence and maintenance of such a cluster of interrelated psychological constructs (e.g., hate or fear of diversity, denial of climate change).

A few circular structures have been discussed in psychology; they include multidimensional models of affect and emotion (Barrett & Russell, 1998, 1999), goals (Grouzet et al., 2005), music functions (Boer et al., 2012; Boer & Fischer, 2012), achievement motivation (Elliot, 2006; Elliot & Thrash, 2002), and vocational interests (Holland, 1996; Tracey & Rounds, 1993). It would be enlightening to examine how these various systems relate to each other. Are there larger organizing principles at a neural, cognitive, social, or environmental level that link such various circular structures together? The human brain may process different types of information in similar ways, which could well lead to a convergence in the psychological organization of all these variables. Similarly, the physical, social, and economic conditions of our environment may make certain structures more likely to emerge. As I already discussed in several places in the book, economic conditions also affect the organization of values and traits. The larger the number of choices that can be afforded with more economic opportunities, the more that human psychology might be reorganized in more profound ways than we typically appreciate.

It is very important that there be truly interdisciplinary work that links neuroscience and genetic research with sociological, economic, and anthropological examinations of external conditions that may affect how we process and integrate information. However, we should be careful about simplifying such relationships too much. Obviously, biological processes are complex and interrelated, even though this complexity is often not apparent when examining some of the research making headlines in academia and popular media. For example, much of the theorizing in the personality literature has focused on serotonin and dopamine, often centering on specific single polymorphisms. More recently, oxytocin has become a neurotransmitter of interest, with a number of studies claiming its effects on a host of personality variables (see Chapter 6). Yet, it is important to reemphasize that many important questions still remain about the biological pathways underlying many of these findings, as well as concerns about the replicability of these findings (Churchland & Winkielman, 2012; Lane, Luminet, Nave, & Mikolajczak, 2016; Nave, Camerer, & McCullough, 2015). Oxytocin is likely to play some role in the larger orchestra of neurotransmitters and neuropeptides. Yet, a single gene that regulates only one small part of the larger process is unlikely to produce complex phenotypes such as personality. More likely, various transmitters, receptors, and promoter genes of multiple neurotransmitters interact, and it is these epistatic effects that play a major role, rather than the single gene–single neurotransmitter–single trait linkages. Research so far has provided relatively little replicable evidence of simple interactions between limited number of genes. It might be more fruitful to consider a systems-gene approach, examining the joint and interactive contribution of genes involved in the regulation of sets of neurotransmitters and neuropeptides (Chen et al., 2011, 2012). We need more research that looks at the interplay of the various neurotransmitters and how they affect the relevant neural processes that are instantiated as value and trait responses. I recommend expanding the gene systems view (similar to the work by Chen and colleagues and also some of the studies that I reviewed in Chapter 8) to focus on larger interacting patterns.

This increased focus on the brain and genes needs to be accompanied by greater sophistication about the social, economic, and ecological context. I have stressed repeatedly the importance of the environment within which individuals are operating. These environmental conditions have a fundamental influence on psychological realities as measured by values and traits. There should no longer be a debate about whether genes or the environment is more important. A more fruitful discussion that we need to have is about how biology and environment jointly give rise to the complex phenotypes of personality. Note that I used terminology from genetics here – but with the explicit inclusion of phenotypic plasticity as a core feature of a personality phenotype. Genes at best predict 50% of the variability in traits and values (if not less), so the environment has a significant role to play in what a person is like. The larger cultural environment seems to explain less than 15% of the variance in most self-report responses. Culture is a large unpackaged construct and needs to be broken down into relevant elements (Poortinga, 2015). Efforts to determine which aspect of the environment is important for which type of personality development deserve more attention. As we saw in Chapter 8, broad macro-contextual features of the environment such as climatic demands or wealth do seem to be important for trait and value expressions. What we do not understand is what specific aspect of climate and wealth (or any other environmental feature) is important and how such features are processed by the organism and lead to the specific personality phenotype. This is a true frontier for further research.

IMPLICATIONS BEYOND ACADEMIA

The core prediction at the theoretical level is that there are a limited number of central motivational systems at the biological level (e.g., behavioral approach vs. avoidance, cooperation) that enable the individual to differentiate whether environmental conditions are supportive and provide opportunities or if the environment is threatening and will limit options. Together with the extension or constriction of the psychological space, the relative importance of traits and values will shift in response to the environment as the person strives toward the end-goal of increasing adaptiveness to the local environment. Individuals socialized into environments where resources are scarce and opportunities for self-expression and personal growth are limited will respond differently to many environmental stimuli compared to individuals who have not faced such economic shortages and are used to seeing environmental challenges as opportunities for self-expression and growth. These differences are likely to have implications across different life contexts; next I will outline a few areas where this theoretical thinking may help us identify and address problems.

Personality and Motivation in the Workplace

A majority of adults in modern societies spend a significant amount of time at work. As we saw in Chapter 9, work and the requirements of specific roles and occupations are likely to shape the personality of individuals. Hence, organizations have some largely unacknowledged role in shaping people and ultimately what type of citizens inhabit a society. Employees entering a workplace that provides economic security, sufficient freedom, and also safety to explore new ideas and use their creative potential will become different people compared with employees entering with short-term contracts, with no financial or job security, working under strict supervision, and with constant threats of punishment for violating policies or procedures.

At the same time, the core mechanisms that are likely to underlie personality dynamics are motivational in nature, which dovetails with a longstanding interest among business managers and scholars in work motivation: how can employees be motivated to work more effectively. Maslow's (1943) hierarchy of needs is often cited (even though it is often more preached than practiced) in this context. Its basic idea is that individuals are only able to express their creative potential if their basic needs are met. The basic tenets of this theory have been supported, although there are questions about the motivational structure of higher-level needs (in particular, the self-actualization needs) positioned at the top of the pyramid (Kenrick, Griskevicius, Neuberg, & Schaller, 2010).

The degree to which a person's basic needs were met during crucial times of development is likely to influence his or her personality structure, which then has implications for how this individual may engage with the work environment. A work environment that provides ample freedom but no structure is unlikely to provide the same motivational value for individuals who have not previously experienced such environments as for those who grew up in a safe and nurturing environment. In threatening and resource-limited environmental contexts, employees may use creative thinking to fix problems and solve immediate crises that help them meet their survival needs, but these opportunities are less likely to be perceived as opportunities for growth and learning if the existential threats are not clearly addressed. Self-centered survival will win over any other motivation, leading to less cooperation, less creativity, less innovation, less open communication, more conformity and limited cooperation to achieve immediate short-term goals of individuals, which ultimately will reduce the competitive advantages of an organization.

To overcome these limitations, managers will need to invest in teams and structures to create a nurturing and safe working environment that also provides the economic security required to overcome an environmentally conditioned avoidance motivation (Fay & Frese, 2001; Frese, Kring, Soose, & Zempel, 1996). Of course not everyone who came from a disadvantaged economic environment will be distrusting and will avoid addressing long-term problems (this is where the phenotypic plasticity argument is central again). The main point here is that work organizations have an impact on personality development that is largely unacknowledged, but has potentially far-reaching consequences beyond the immediate work environment. By investing in creating safe and supportive work environments, organizations can reap huge benefits by developing a creative, cooperative, and motivated workforce, as well as having a positive impact on the larger society. It remains to be seen to what extent such initiatives find traction in the current political and economic climate, where short-term economic gains trump long-term growth.

Mental Health

Clinical syndromes are often seen as separate and distinct from population traits in the general population. Equally, personality disorders are often seen as discrete types that are unconnected to personality characteristics as discussed in this book. Yet, the emerging evidence indicates that clinical problems are extensions of normal personality variation into what is deemed problematic from a "normative" perspective. For example, genes that are implicated in personality disorders are also involved in trait phenotypes in the general population (Lo et al., 2016). Personality traits and values are shaped by genes and become expressed in a particular environment. From a fitness perspective, individuals will maximize their chances of survival and reproduction, which then leads to certain behavioral and cognitive responses to environmental cues. However, there are many possibilities for mismatch. Early experiences during infancy that signal one type of environment may be a mismatch for conditions later in the life of individuals. Infants growing up in insecure and resource-limited environments may develop aggressive, manipulative, or antisocial traits and values (shifting toward instrumental trust orientations and extreme forms of either behavioral approach or avoidance motivation) that would benefit them in such an environment. These tendencies may then appear dysfunctional when considered in a larger context that is not characterized by the same environmental conditions (Belsky & Pluess, 2013; Ellis et al., 2011). The growing research on gene-environment interactions also provides novel insights into those individuals who are both more vulnerable when faced with difficult life situations and more likely to thrive when encountering nurturing and supportive environments (Bakermans-Kranenburg & van Ijzendoorn, 2011). This significantly broadens the pathologically focused perspectives and removes stigma from individuals who might otherwise be labeled as weak or instable. These insights can be used to help them overcome challenges and reach their full potential as individuals.

Equally, how values relate to well-being and depression depends to at least some extent on the larger environment and what other people value. Florencia Sortheix and her colleagues have demonstrated in a number of studies that the larger social and economic context matters when talking about values and life satisfaction (Sortheix & Lönnqvist, 2014, 2015; Sortheix & Schwartz, 2017). When our values match those of most others in our society, we then experience greater life satisfaction. If we are concerned about the well-being of our loved ones and worry about conforming to social standards (Security and Conformity values, among others), but happen to live in an environment that is oppressive and characterized by poverty, we are more likely to experience low life satisfaction and to develop depression.

Mental health, clinical, and psychiatric research and practice are highly diverse in orientation and emphasis. What is striking, however, is the separation of the clinical field from other areas of psychology, biology, and anthropology that may provide new insights into clinical practice and training. For example, the recognition that clinical symptoms and mental health problems might be adaptations to the environment that became mismatched sometime during the development of an individual may open up interesting opportunities for personal healing and growth. Realigning values and traits with current environmental conditions through clinical interventions could help individuals overcome problems and help them reintegrate into society (Ward & Gannon, 2006; Ward, Mann, & Gannon, 2007).

Immigration and Acculturation

Probably one of the biggest social and political issues in the last few years has been the public debate over immigration and acculturation – how people should and could adjust to their new host society. The issue was at the forefront of such seminal events as the vote of British citizens to leave the European Union and the election of Donald Trump to become the US president. Both the individuals who migrate to a new place and the individuals who receive them as hosts are shaped by migration. What are the implications for personality?

One of the issues raised especially in Europe is that migrants from non-European countries have not integrated into the host culture and do not participate fully in civic life. It is important to realize that many of the migrants enter their new host country as political or economic refugees. They are escaping a home that is often characterized by extreme poverty and war; they lack opportunities to even satisfy basic needs or find meaningful and sufficiently renumerative work that would enable them and their families to thrive. As discussed throughout this book, these conditions are likely to foster a particular personality profile, one that is likely to be adaptive in these extreme social and economic environments, but that might mismatch the requirements in their new host environment. Exacerbating these conditions even more, most migrants and refugees continue to face economic disadvantages after they enter their new host environment. For example, refugees in Europe are congregated in camp-like structures, are often not allowed to work, and, if they do work, are often employed in the lowest level jobs available within the economy. There are limited opportunities for migrants to adjust to the new environment and develop the traits and values that are deemed appropriate from the hosts' perspective. Refugees and migrants may aspire to achieve the lofty goals of acculturation and integration into a well-functioning civil society, but the reality they face of exclusion and continuing deprivation constrains skill development and achievements toward that end.

Participating in democratic institutions and adopting new symbolic customs and values are only possible if economic needs are met and migrants feel safe to explore these new freedoms. What are the incentives to participate if people are consumed by worry about the economic situation and remain marginalized? In the study of great tits that I discussed in the previous chapter (Dingemanse et al., 2004), competition is beneficial if there are limited resources. Aggressive and antisocial behavior is adaptive in these conditions, and people fleeing war zones and extreme poverty are probably well advised to behave in this way. Yet, in a new environment, other behavioral traits and values are adaptive. Migrants and refugees have to be given the space to adapt to these new environments and to adapt their traits and values by taking up new roles. As the work on organizational socialization and clinical interventions has shown, personality is likely to change and has quite high plasticity even late in life. But this change requires time and patience on behalf of all people involved. To expect individuals to change their motivational orientation within a short period of time of weeks or months (without often having the appropriate conditions to do so) is probably naïve.

At a more fundamental level, the **motivational differentiation** hypothesis discussed in Chapter 6 implies that even the opportunities that are present for migrants might not be readily recognized because previous environmental conditions have led to a different personality structure that might have been adaptive in that environment. For example, in our democratic environments, taking care of one's family and close friends often involves participating in community activities, speaking up about proposed policy changes, or even taking part in a peaceful demonstration or protest. In contrast, individuals who have not had opportunities during their formative years to develop a well-differentiated set of values separating growth and protection values in Schwartz's terminology might support actions that might seem contradictory or perplexing from an outsider perspective. Their actions are guided by their experiences, which may be a mismatch to the new environment. Therefore, the very participation that is desired by the hosts might lead to rejection or misunderstanding because it is based on a different set of worldviews corresponding to conditions in different economic and social realities. I am not arguing against universal human rights here. On the contrary, I argue that we all have the same fundamental human rights and the same potential to pursue these rights. Yet, whether and how we have experienced opportunities to express these rights differ across political and economic systems in today's world, and this divergence of human experience that shaped expectations and (in)actions of individuals and communities needs more appreciation.

Switching the lens to the hosts, it is understandable that they have frustrations and anxieties. The influx of migrants into more economically advanced societies the world over is often perceived as increasing economic insecurity ("migrants are taking our jobs") and perceptions that migrants and refugees cause violence and crime. These perceptions are probably inaccurate (e.g., migrants often stimulate economic growth in advanced industries with diminishing birth rates; migrants in the United States are less likely to commit a crime compared to US citizens). Nevertheless, these fears increase behavioral avoidance motivations and instrumental cooperation orientations among hosts that then further increase their fears. As a consequence, through the aggregation of these processes happening at the individual level, we see a conservative shift and radicalization of social groups that feel most threatened. Thus economic and social security needs of the hosts need to be acknowledged and addressed. The basics are quite simple: people need to feel secure about their place in society, including economic and social conditions. Unfortunately, the political will to address these issues is often missing.

Economic Development

One of the clear messages is that economic conditions do matter for individuals and profoundly shape personality, both traits and values. Many people might be aghast at these findings and perceive them as being part of a neoliberal agenda. I think this is far from the truth. The issue is that economic conditions need to meet basic needs of all people, and therefore, economic development that is built around social fairness is crucial. Individuals need to be given the economic means to address challenges in their environment and to learn to act on the opportunities there. Without addressing these needs first, little personal growth or development in the way that is desired by social activists might be achievable (on average – there are always examples to the contrary). This view clashes with those popular among participatory action researchers and activists who promote grassroots action as a primary tool to empower and transform communities (e.g., Baum, McDougall, & Smith, 2006; Ratner, 2009; Williams, Labonte, & O'Brien, 2003). It is questionable whether such action-focused initiatives can be successful if the motivational orientations of participants are constrained by economic deprivation (or adverse environmental conditions more generally). Individuals and communities need to be given (a) economic resources to transform and improve economic conditions, and (b) communities need to be assisted and supported in their collaborative effort for long-term growth and human development.

A singular emphasis on economics is nevertheless not a panacea. For example, micro-crediting is a popular economic development initiative in much of the majority world. But giving individuals money without providing the infrastructure for them to develop is unlikely to bring the desired effects, at least in the short term (Karlan & Zinman, 2011). Again, a holistic approach informed by insights into human development and that considers the adaptiveness of the personality structure of individuals within a given environment is required to determine what other support is needed in addition to money. Evaluation of interventions is also important, but it is important to realize that time frames for effects to be reliably picked up might be medium to long term. Overall, a balanced approach is needed that takes into account multidisciplinary perspectives on human development.

SUMMARY

I started off with a colorful example of human diversity evident in contemporary societies. It is easy to focus on the apparent exotic practices and quickly jump to conclusions about cultural differences in values, traits, and broader personality dimensions. Tina and Sandra in their behaviors appear fundamentally different, and it is hard to imagine how they might have any similar underlying values or personality traits. One of my aims was to examine the extent to which whether values and traits are different and how any such differences could be understood from an evolutionary perspective.

Focusing on data from studies around the world, we realize that there are actually relatively small differences in the underlying psychological make-up of humans. The variability that we see between groups is far exceeded by the differences within groups. Yet, the mechanisms that may explain differences between individuals within groups also provide insights into why groups of people living in different parts of the world show slightly different personality characteristics. I have argued throughout the book that the examination of evolutionary constraints on adaptation and development is an informative approach for understanding personality. Tina and Sandra after all are humans like you and me. However, both these women, their families, and their ancestors certainly have encountered different living conditions that gave rise to these diverse behaviors that I described in the opening paragraphs of this book. There is nothing exotic about this: it is part and parcel of who we are as a biological species. We need to look beyond the exotic nature of these behaviors and examine the underlying dynamics.

My hope is that the theoretical approach presented here will inspire new research that will bridge disciplines and research traditions. I combined two diverse strands of research (values and traits) and linked them to the same underlying biological mechanisms. Other constructs in psychology might be organized via similar processes. At the same, I tried to avoid biological reductionism by stressing the importance of the social, economic, and ecological conditions within which people operate. The discussion should not be about biology versus culture, but rather about biology and culture together. But culture should be dissected, the same way that biologists have examined different genes, neurotransmitters, and ecological conditions. This theoretical model is a step in this direction, a gross over-simplification for sure, but a starting point that I hope will inspire others to go beyond. I also hope I have convinced you that this is an exciting field of research with many possible applications and implications that are worth considering in the real world.

REFERENCES

Aavik, T., & Allik, J. (2002). The structure of Estonian personal values: A lexical approach. *European Journal of Personality, 16*(3), 221–235. DOI:10.1002/per.439

Adelstein, J., Shehzad, Z., Mennes, M., DeYoung, C., Zuo, X.-N., Kelly, C., ... Milham, M. (2011). Personality is reflected in the brain's intrinsic functional architecture. *PLoS ONE, 6*(11), e27633. DOI:10.1371/journal.pone.0027633

Allen, T., & DeYoung, C. (2017). Personality neuroscience and the five factor model. In T. A. Widiger (ed), *Oxford handbook of the five factor model*. Oxford: Oxford University Press. DOI:10.1093/oxfordhb/9780199352487.013.26

Allik, J., & McCrae, R. (2004). Toward a geography of personality traits: Patterns of profiles across 36 cultures. *Journal of Cross-Cultural Psychology, 35*(1), 13–28. DOI:10.1177/0022022103260382

Allport, G. (1955). *Becoming; Basic considerations for a psychology of personality*. New Haven, CT: Yale University Press.

Allport, G., & Odbert, H. (1936). Trait-names: A psycho-lexical study. *Psychological Monographs, 47*, 1–171.

Alvergne, A., Jokela, M., & Lummaa, V. (2010). Personality and reproductive success in a high-fertility human population. *Proceedings of the National Academy of Sciences, 107*(26), 11745–11750. DOI:10.1073/pnas.1001752107

Ashton, M., & Lee, K. (2007). Empirical, theoretical, and practical advantages of the HEXACO model of personality structure. *Personality and Social Psychology Review, 11*, 150–166.

Ashton, M., & Lee, K. (2008). The prediction of honesty-humility-related criteria by the HEXACO and Five-Factor models of personality. *Journal of Research in Personality, 42*, 1216–1228.

Ashton, M., Lee, K., & Goldberg, L. (2004). A hierarchical analysis of 1,710 English personality-descriptive adjectives. *Journal of Personality and Social Psychology, 87*(5), 707. DOI:10.1037/0022-3514.87.5.707

Ashton, M., Lee, K., Goldberg, L., & de Vries, R. (2009). Higher order factors of personality: Do they exist? *Personality and Social Psychology Review, 13*, 79–91.

Bailey, D., Walker, R., Blomquist, G., Hill, K., Hurtado, M., & Geary, D. (2013). Heritability and fitness correlates of personality in the Ache, a natural-fertility population in Paraguay. *PLoS ONE, 8*(3), e59325. DOI:10.1371/journal.pone.0059325

Bakermans-Kranenburg, M., & van Ijzendoorn, M. (2011). Differential susceptibility to rearing environment depending on dopamine-related genes: New evidence and a meta-analysis. *Development and Psychopathology, 23*(01), 39–52. DOI:10.1017/s0954579410000635

Baltes, P. (1987). Theoretical propositions of life-span developmental psychology: On the dynamics between growth and decline. *Developmental Psychology, 23*, 611–626.

Baltes, P., & Lindenberger, U. (1997). Emergence of a powerful connection between sensory and cognitive functions across the adult life span: A new window to the study of cognitive aging? *Psychology and Aging, 12*(1), 12. DOI:10.1037/0882-7974.12.1.12

Baltes, P., Reese, H. W., & Lipsitt, L. P. (1980). Life-span developmental psychology. *Annual Review of Psychology, 31*(1), 65–110. DOI:10.1146/annurev.ps.31.020180.000433

Baltes, P., & Schaie, W. (1973). *Life-span developmental psychology: Personality and socialization*. London: Academic Press.

Baltes, P. B., Staudinger, U. M., & Lindenberger, U. (1999). Lifespan psychology: Theory and application to intellectual functioning. *Annual Review of Psychology, 50*(1), 471–507. DOI:10.1146/annurev.psych.50.1.471

Barr, R. G., Konner, M., Bakeman, R. & Adamson, L. (1991). Crying in! Kung San infants: A test of the cultural specificity hypothesis. *Developmental Medicine and Child Neurology*, 33, 601–610. DOI:10.1111/j.1469-8749.1991.tb14930.x

Barrett, L., & Russell, J. (1998). Independence and bipolarity in the structure of current affect. *Journal of Personality and Social Psychology, 74*(4), 967–984.

Barrett, L., & Russell, J. (1999). The structure of current affect. *Current Directions in Psychological Science, 8*(1), 10–14. DOI:10.1111/1467-8721.00003

Barrett, P. T., Petrides, K. V., Eysenck, S. B. G., & Eysenck, H. J. (1998). The Eysenck Personality Questionnaire: An examination of the factorial similarity of P, E, N, and L across 34 countries. *Personality and Individual Differences, 25*(5), 805–819. DOI:10.1016/s0191-8869(98)00026-9

Bartram, D. (2013). A cross-validation of between country differences in personality using the OPQ32. *International Journal of Quantitative Research in Education, 1*(2), 182. DOI:10.1504/ijqre.2013.056460

Bartram, D., & Brown, A. (2005). *Five Factor Model (Big Five) OPQ32 report: OPQ32 technical manual supplement*. Thames Ditton, UK: SHL Group.

Baum, F., MacDougall, C., & Smith, D. (2006). Participatory action research. *Journal of Epidemiology and Community Health, 60*(10), 854–857. DOI:10.1136/jech.2004.028662

Beckmann, N., Wood, R., & Minbashian, A. (2010). It depends how you look at it: On the relationship between neuroticism and conscientiousness at the within- and the between-person levels of analysis. *Journal of Research in Personality, 44*(5), 593–601. DOI:10.1016/j.jrp.2010.07.004

Beja-Pereira, A., Luikart, G., England, P., Bradley, D., Jann, O., Bertorelle, G., . . . Erhardt, G. (2003). Gene-culture coevolution between cattle milk protein genes and human lactase genes. *Nature Genetics*, *35*(4), 311–313. DOI:10.1038/ng1263

Belsky, J., & Pluess, M. (2013). Beyond risk, resilience, and dysregulation: Phenotypic plasticity and human development. *Development and Psychopathology*, 25(4, pt. 2), 1243–1261. DOI:10.1017/s095457941300059x

Bem, D. (1967). Self-perception: An alternative interpretation of cognitive dissonance phenomena. *Psychological Review*, *74*(3), 183–200. DOI:10.1037/h0024835

Bem, D. (1972). Self-perception theory. In L. Berkowitz (Ed.), *Advances in experimental social psychology*, vol. 6, pp. 1–62. New York: Academic Press.

Bender, M., & Chasiotis, A. (2011). Number of siblings in childhood explains cultural variance in autobiographical memory in Cameroon, People's Republic of China, and Germany. *Journal of Cross-Cultural Psychology*, *2*(6), 998–1017.

Benet-Martínez, V., & John, O. (1998). Los Cinco Grandes across cultures and ethnic groups: Multitrait-multimethod analyses of the Big Five in Spanish and English. *Journal of Personality and Social Psychology*, *75*(3), 729–750. DOI:10.1037//0022-3514.75.3.729

Benjamin, J., Li, J., Patterson, C., Greenberg, B. D., Murphy, D. L., & Hamer, D. H. (1996). Population and familial association between the D4 dopamine receptor gene and measures of Novelty Seeking. *Nature Genetics*, *12*(1), 81–84.

Berridge, K. C. (1996). Food reward: Brain substrates of wanting and liking. *Neuroscience and Biobehavioral Reviews*, *20*(1), 1–25.

Berridge, K. C., Robinson, T. E., & Aldridge, J. W. (2009). Dissecting components of reward: "liking," "wanting," and learning. *Current Opinion in Pharmacology*, 9(1), 65–73. DOI:10.1016/j.coph.2008.12.014

Berry, J., Poortinga, Y., Breugelmans, S., Chasiotis, A., & Sam, D. (2011). *Cross-cultural psychology: Research and applications*. Cambridge: Cambridge University Press.

Berson, Y., & Oreg, S. (2016). The role of school principals in shaping children's values. *Psychological Science*, *27*(12), 1539–1549. DOI:10.1177/0956797616670147

Bijma, P. (2014). The quantitative genetics of indirect genetic effects: A selective review of modelling issues. *Heredity*, *112*(1), 61–69. DOI:10.1038/hdy.2013.15

Bilsky, W., Janik, M., & Schwartz, S. H. (2011). The structural organization of human values – evidence from three rounds of the European Social Survey (ESS). *Journal of Cross-Cultural Psychology*, *42*(5), 759–776. DOI:10.1177/0022022110362757

Bleidorn, W., Klimstra, T., Denissen, J., Rentfrow, P., Potter, J., & Gosling, S. (2013). Personality maturation around the world. *Psychological Science*, *24*(12), 2530–2540. DOI:10.1177/0956797613498396

Boer, D., & Fischer, R. (2012). Towards a holistic model of functions of music listening across cultures: A culturally decentred qualitative approach. *Psychology of Music*, *40*(2), 179–200. DOI:10.1177/0305735610381885

Boer, D., & Fischer, R. (2013). How and when do personal values guide our attitudes and sociality? Explaining cross-cultural variability in attitude–value linkages. *Psychological Bulletin*, *139*(5), 1113–1147. DOI:10.1037/a0031347

Boer, D., Fischer, R., Tekman, H., Abubakar, A., Njenga, J., & Zenger, M. (2012). Young people's topography of musical functions: Personal, social and cultural experiences with music across genders and six societies. *International Journal of Psychology, 47*(5), 355–369. DOI:10.1080/00207594.2012.656128

Bond, M. (1988). Finding universal dimensions of individual variation in multicultural studies of values: The Rokeach and Chinese value surveys. *Journal of Personality and Social Psychology, 55*(6), 1009–1015. DOI:10.1037//0022-3514.55.6.1009

Bonifay, W., Lane, S. P., & Reise, SP. (2017). Three concerns with applying a bifactor model as a structure of psychopathology. *Clinical Psychological Science, 5*, 184–186.

Borg, I. & Groenen, P. (2003). Modern multidimensional scaling: Theory and applications. *Journal of Educational Measurement, 40*(3), 277–280. DOI:10.1111/j.1745-3984.2003.tb01108.x

Borg, I., & Groenen, P. (2005). *Modern multidimensional scaling: Theory and applications* (2nd ed.). New York: Springer.

Borsboom, D. (2006). The attack of the psychometricians. *Psychometrika, 71*(3), 425–440. DOI:10.1007/s11336-006-1447-6

Borsboom, D. (2008). Latent variable theory. *Measurement: Interdisciplinary Research & Perspective, 6*(1–2), 25–53. DOI:10.1080/15366360802035497

Borsboom, D., Mellenbergh, G., & van Heerden, J. (2003). The theoretical status of latent variables. *Psychological Review, 110*(2), 203–219. DOI:10.1037/0033-295X.110.2.203

Bouchard, T. (2004). Genetic influence on human psychological traits: A survey. *Current Directions in Psychological Science, 13*, 148–151.

Brosch, T., Coppin, G., Scherer, K., Schwartz, S., & Sander, D. (2011). Generating value(s): Psychological value hierarchies reflect context-dependent sensitivity of the reward system. *Social Neuroscience, 6*(2), 198–208. DOI:10.1080/17470919.2010.506754

Brosch, T., & Sander, D. (2013). Neurocognitive mechanisms underlying value-based decision-making: From core values to economic value. *Frontiers in Human Neuroscience, 7*(398), 1–8. DOI:10.3389/fnhum.2013.00398

Brown, G., Dickins, T., Sear, R., & Laland, K. (2011). Evolutionary accounts of human behavioural diversity. *Philosophical Transactions of the Royal Society B: Biological Sciences, 366*(1563), 313–324. DOI:10.1098/rstb.2010.0267

Buss, D. (1991). Evolutionary personality psychology. *Annual Review of Psychology, 42*(1), 459–491. DOI:10.1146/annurev.ps.42.020191.002331

Buss, D. (2001). Human nature and culture: An evolutionary psychological perspective. *Journal of Personality, 69*(6), 955–978. DOI:10.1111/1467-6494.696171

Buss, D. (2009). How can evolutionary psychology successfully explain personality and individual differences? *Perspectives on Psychological Science, 4*(4), 359–366. DOI:10.1111/j.1745-6924.2009.01138.x

Carver, C., & Scheier, M. (1982). Control theory: A useful conceptual framework for personality-social, clinical, and health psychology. *Psychological Bulletin, 92*(1), 111–135.

Carver, C., & Scheier, M. (2002). Control processes and self-organization as complementary principles underlying behavior. *Personality and Social Psychology Review, 6*, 304–315.

Caspi, A., Harrington, H., Milne, B., Amell, J., Theodore, R., & Moffitt, T. (2003). Children's behavioral styles at age 3 are linked to their adult personality traits at age 26. *Journal of Personality*, *71*(4), 495–514. DOI:10.1111/1467-6494.7104001

Chabris, C., Lee, J., Cesarini, D., Benjamin, D., & Laibson, D. (2015). The fourth law of behavior genetics. *Current Directions in Psychological Science*, *24*(4), 304–312.

Champagne, F. (2008). Epigenetic mechanisms and the transgenerational effects of maternal care. *Frontiers in Neuroendocrinology*, *29*(3), 386–397. DOI:10.1016/j.yfrne.2008.03.003

Chen, C., Chen, C., Moyzis, R., He, Q., Li, H., Li, J., . . . Dong, Q. (2012). Genetic variations in the dopaminergic system and alcohol use: A system-level analysis. *Addiction Biology*, *17*(2), 479–489. DOI:10.1111/j.1369-1600.2011.00348.x

Chen, C., Chen, C., Moyzis, R., Stern, H., He, Q., Li, H., . . . Dong, Q. (2011). Contributions of dopamine-related genes and environmental factors to highly sensitive personality: A multi-step neuronal system-level approach. *PloS one*, *6*(7), e21636. DOI:10.1371/journal.pone.0021636

Chen, C., Moyzis, R., Lei, X., Chen, C., & Dong, Q. (2016). The encultured genome: Molecular evidence for recent divergent evolution in human neurotransmitter genes. In J. Chiao, S. Li, R. Seligman, & R. Turner (Eds.), *The Oxford handbook of cultural neuroscience* (pp. 315–336). Oxford: Oxford University Press.

Cheung, F., Fan, W., & To, C. (2008). The Chinese Personality Assessment Inventory as a culturally relevant personality measure in applied settings. *Social and Personality Psychology Compass*, *2*(1), 74–89. DOI:10.1111/j.1751-9004.2007.00045.x

Cheung, F., Leung, K., Fan, R., Song, W.-Z., Zhang, J.-X., & Zhang, J.-P. (1996). Development of the Chinese Personality Assessment Inventory. *Journal of Cross-Cultural Psychology*, *27*(2), 181–199. DOI:10.1177/0022022196272003

Cheung, F., Leung, K., Song, W.-Z., & Zhang, J.-X. (2001). *The cross-cultural (Chinese) Personality Assessment Inventory-2 (CPAI-2)*. Hong Kong: Chinese University of Hong Kong.

Cheung, F., Leung, K., Zhang, J.-X., Sun, H.-A., Gan, Y.-Q., Song, W.–Z, & Xie, D. (2001). Indigenous Chinese personality constructs: Is the Five-Factor Model complete? *Journal of Cross-Cultural Psychology*, *32*(4), 407–433. DOI:10.1177/0022022101032004003

Cheung, F., van de Vijver, F., & Leong, F. (2011). Toward a new approach to the study of personality in culture. *American Psychologist*, *66*(7), 593–603. DOI:10.1037/a0022389

Cheung, S., Cheung, F., Howard, R., & Lim, Y.-H. (2006). Personality across the ethnic divide in Singapore: Are "Chinese Traits" uniquely Chinese? *Personality and Individual Differences*, *41*(3), 467–477. DOI:10.1016/j.paid.2005.12.023

Chiao, J., & Blizinsky, K. (2010). Culture–gene coevolution of individualism–collectivism and the serotonin transporter gene. *Proceedings of the Royal Society B: Biological Sciences*, *277*(1681), 529–537. DOI:10.1098/rspb.2009.1650

Church, T., Alvarez, J., Mai, N., French, B., Katigbak, M., & Ortiz, F. (2011). Are cross-cultural comparisons of personality profiles meaningful? Differential item and facet functioning in the Revised NEO Personality Inventory. *Journal of Personality and Social Psychology*, *101*(5), 1068. DOI:10.1037/a0025290

Churchland, P. S., & Winkielman, P. (2012). Modulating social behavior with oxytocin: How does it work? What does it mean? *Hormones and Behavior*, *61*, 392–399.

Cieciuch, J., Davidov, E., & Algesheimer, R. (2016). The stability and change of value structure and priorities in childhood: A longitudinal study. *Social Development*, 25(3), 503–527. DOI:10.1111/sode.12147

Cloninger, C.R. (1986). A unified biosocial theory of personality and its role in the development of anxiety states. *Psychiatric Developments*, *4*(3), 167–226.

Cloninger, C. R., Przybeck, T. R., & Svrakic, D. M. (1991). The Tridimensional Personality Questionnaire: U.S. normative data. *Psychological Reports*, *69*, 1047–1057.

Cohen, J. (1988). *Statistical power analysis for the behavioral sciences* (2nd ed.) Hillsdale, NJ: Erlbaum.

Cohen, J. (1992). Statistical power analysis. *Current Directions in Psychological Science*, *1*, 98–101.

Collier, D. A., Stöber, G., Li, T., Heils, A., Catalano, M., Di Bella, D., . . . Lesch, K. P. (1996). A novel functional polymorphism within the promoter of the serotonin transporter gene: Possible role in susceptibility to affective disorders. *Molecular Psychiatry*, *1*(6), 453–460.

Corr, P. (2004). Reinforcement sensitivity theory and personality. *Neuroscience and Biobehavioral Reviews*, *28*(3), 317–332. DOI:10.1016/j.neubiorev.2004.01.005

Corr, P. (2009). The reinforcement sensitivity theory of personality. In P. Corr, and G. Matthews (Eds.), *The Cambridge handbook of personality psychology* (pp. 340–346). Cambridge: Cambridge University Press.

Corr, P. J., DeYoung, C. G., & McNaughton, N. (2013). Motivation and personality: A neuropsychological perspective. *Social and Personality Psychology Compass*, *7*(3), 158–175. DOI:10.1111/spc3.12016

Cosmides, L., & Tooby, J. (1989). Evolutionary psychology and the generation of culture, part II. *Ethology and Sociobiology*, *10*(1–3), 51–97. DOI:10.1016/0162-3095(89)90013-7

Cosmides, L., & Tooby, J. (1992). Cognitive adaptations for social exchange. In J. Barkow, L. Cosmides, & J. Tooby (Eds.), *The adapted mind: Evolutionary psychology and the generation of culture* (pp. 163–228). New York: Oxford University Press.

Costa, P. T., & McCrae, R. R. (1992). *Revised NEO Personality Inventory (NEO-PI-R) and NEO Five-Factor Inventory (NEO-FFI) professional manual.* Odessa, FL: Psychological Assessment Resources.

Courtiol, A., Tropf, F., & Mills, M. (2016). When genes and environment disagree: Making sense of trends in recent human evolution. *Proceedings of the National Academy of Sciences*, *113*(28), 7693–7695. DOI:10.1073/pnas.1608532113

Cramer, A., Sluis, S., Noordhof, A., Wichers, M., Geschwind, N., Aggen, S., . . . Borsboom, D. (2012). Dimensions of normal personality as networks in search of equilibrium: You can't like parties if you don't like people. *European Journal of Personality*, *26*(4), 414–431. DOI:10.1002/per.1866

Cuddy, A., Fiske, S., & Glick, P. (2007). The BIAS map: Behaviors from intergroup affect and stereotypes. *Journal of Personality and Social Psychology*, *92*(4), 631–648. DOI:10.1037/0022-3514.92.4.631

Cuddy, A., Fiske, S., & Glick, P. (2008). Warmth and competence as universal dimensions of social perception: The stereotype content model and the BIAS map. *Advances in Experimental Social Psychology* (vol. 40, pp. 61–149). DOI:10.1016/s0065-2601(07)00002-0

Cuddy, A., Fiske, S., Kwan, V., Glick, P., Demoulin, S., Leyens, J., . . . Ziegler, R. (2009). Stereotype content model across cultures: Towards universal similarities and some differences. *British Journal of Social Psychology*, *48*(1), 1–33. DOI:10.1348/014466608x314935

Darwin, C. (1859). *On the origins of species by means of natural selection*. London: Murray.

Dash, S., Clarke, G., Berk, M., & Jacka, F. (2015). The gut microbiome and diet in psychiatry: Focus on depression. *Current Opinion in Psychiatry*, *28*(1), 1–6. DOI:10.1097/yco.0000000000000117

Davidov, E., Schmidt, P., & Schwartz, S. H. (2008). Bringing values back in: The adequacy of the European Social Survey to measure values in 20 countries. *Public Opinion Quarterly*, *72*(3), 420–445. DOI:10.1093/poq/nfn035

Dawkins, R. (1976). *The selfish gene*, Oxford: Oxford University Press.

De Moor, M. H. M., Costa, P. T., Terracciano, A., Krueger, R. F., de Geus, E. J. C., Toshiko, T., . . . Boomsma, D. I. (2010). Meta-analysis of genome-wide association studies for personality. *Molecular Psychiatry*, *17*(3), 337–349. DOI:10.1038/mp.2010.128

De Moor, M. H. M., van den Berg, S., Verweij, K., Krueger, R., Luciano, M., Vasquez, A., . . . Boomsma, D. (2015). Meta-analysis of genome-wide association studies for Neuroticism, and the polygenic association with Major Depressive Disorder. *JAMA Psychiatry*, *72*(7), 642–650. DOI:10.1001/jamapsychiatry.2015.0554

De Raad, B., Barelds, D. P., Levert, E., Ostendorf, F., Mlačić, B., Blas, L., . . . Katigbak, M. S. (2010). Only three factors of personality description are fully replicable across languages: A comparison of 14 trait taxonomies. *Journal of Personality and Social Psychology*, *98*(1), 160–173. DOI:10.1037/a0017184

De Raad, B., Barelds, D. P., Timmerman, M. E., Roover, K., Mlačić, B., & Church, T. A. (2014). Towards a pan-cultural personality structure: Input from 11 psycholexical studies. *European Journal of Personality*, *28*(5), 497–510. DOI:10.1002/per.1953

De Raad, B., & Oudenhoven, J. (2008). Factors of values in the Dutch language and their relationship to factors of personality. *European Journal of Personality*, 22(2), 81–108. DOI:10.1002/per.667

Declerck, C., Boone, C., & Emonds, G. (2013). When do people cooperate? The neuroeconomics of prosocial decision making. *Brain and Cognition*, *81*(1), 95–117. DOI:10.1016/j.bandc.2012.09.009

DeYoung, C. (2006). Higher-order factors of the Big Five in a multi-informant sample. *Journal of Personality and Social Psychology*, *91*(6), 1138–1151. DOI:10.1037/0022-3514.91.6.1138

DeYoung, C. (2010). Personality neuroscience and the biology of traits. *Social and Personality Psychology Compass*, 4(12), 1165–1180. DOI:10.1111/j.1751-9004.2010.00327.x

DeYoung, C. (2013). The neuromodulator of exploration: A unifying theory of the role of dopamine in personality. *Frontiers in Human Neuroscience*, *7*. DOI:10.3389/fnhum.2013.00762

DeYoung, C. (2014). Cybernetic Big Five theory. *Journal of Research in Personality, 56*, 33–58. DOI:10.1016/j.jrp.2014.07.004

DeYoung, C., Peterson, J., & Higgins, D. (2002). Higher-order factors of the Big Five predict conformity: Are there neuroses of health? *Personality and Individual Differences, 33*(4), 533–552. DOI:10.1016/s0191-8869(01)00171-4

DeYoung, C., Quilty, L., & Peterson, J. (2007). Between facets and domains: 10 aspects of the Big Five. *Journal of Personality and Social Psychology, 93*(5), 880. DOI:10.1037/0022-3514.93.5.880

Digman, J. (1997). Higher-order factors of the Big Five. *Journal of Personality and Social Psychology, 73*(6), 1246–1256. DOI:10.1037/0022-3514.73.6.1246

Dingemanse, N. J., Both, C., Drent, P. J., & Tinbergen, J. M. (2004). Fitness consequences of avian personalities in a fluctuating environment. *Proceedings of the Royal Society of London. Series B: Biological Sciences, 271*(1541), 847–852. DOI:10.1098/rspb.2004.2680

Dingemanse, N. J., Kazem, A., Réale, D., & Wright, J. (2010). Behavioural reaction norms: Animal personality meets individual plasticity. *Trends in Ecology & Evolution, 25*(2), 81–89. DOI:10.1016/j.tree.2009.07.013

Dobewall, H., & Rudnev, M. (2014). Common and unique features of Schwartz's and Inglehart's value theories at the country and individual levels. *Cross-Cultural Research, 48*, 45–77.

Donnellan, B., Oswald, F., Baird, B., & Lucas, R. (2006). The Mini-IPIP Scales: Tiny-yet-effective measures of the Big Five Factors of Personality. *Psychological Assessment, 18*(2), 192. DOI:10.1037/1040-3590.18.2.192

Döring, A., Schwartz, S., Cieciuch, J., Groenen, P., Glatzel, V., Harasimczuk, J., . . . Bilsky, W. (2015). Cross-cultural evidence of value structures and priorities in childhood. *British Journal of Psychology, 106*(4), 675–699. DOI:10.1111/bjop.12116

Dunkel, C., Mathes, E., Kesselring, S., Decker, M., & Kelts, D. (2015). Parenting influence on the development of life history strategy. *Evolution and Human Behavior, 36*, 374–378.

Durham, W. (1991). *Coevolution: Genes, culture, and human diversity.* Stanford: Stanford University Press.

Ebstein, R. P., Knafo, A., Mankuta, D., & Chew, S. H. (2012). The contributions of oxytocin and vasopressin pathway genes to human behavior. *Hormones and Behavior, 61*, 359–379.

Elliot, A. (2006). The hierarchical model of approach-avoidance motivation. *Motivation and Emotion, 30*(2), 111–116. DOI:10.1007/s11031-006-9028-7

Elliot, A., & Thrash, T. (2002). Approach-avoidance motivation in personality: Approach and avoidance temperaments and goals. *Journal of Personality and Social Psychology, 82*(5), 804–818. DOI:10.1037//0022-3514.82.5.804

Ellis, B., Boyce, T., Belsky, J., Bakermans-Kranenburg, M., & van Ijzendoorn, M. (2011). Differential susceptibility to the environment: An evolutionary–neurodevelopmental theory. *Development and Psychopathology, 23*(1), 7–28. DOI:10.1017/s0954579410000611

Eppig, C., Fincher, C., & Thornhill, R. (2010). Parasite prevalence and the worldwide distribution of cognitive ability. *Proceedings of the Royal Society B: Biological Sciences, 277*(1701), 3801–3808. DOI:10.1098/rspb.2010.0973

Erikson, E., & Erikson, J. (1998). *The life cycle completed* (extended version). New York: Norton.

Eysenck, H., & Eysenck, S. (1965). The Eysenck personality inventory. *British Journal of Educational Studies, 14*, 140.

Eysenck, H., & Eysenck, S. (1975). *Manual of the Eysenck Personality Questionnaire (junior and adult)*. London: Hodder and Stoughton.

Fay, D., & Frese, M. (2001). The concept of personal initiative: An overview of validity studies. *Human Performance, 14*(1), 97–124. DOI:10.1207/s15327043hup1401_06

Fazio, R., Zanna, M., & Cooper, J. (1977). Dissonance and self-perception: An integrative view of each theory's proper domain of application. *Journal of Experimental Social Psychology, 13*, 464–479.

Feldman, R., Monakhov, M., Pratt, M., & Ebstein, R. (2016). Oxytocin pathway genes: Evolutionary ancient system impacting on human affiliation, sociality, and psychopathology. *Biological Psychiatry, 79*(3), 174–184. DOI:10.1016/j.biopsych.2015.08.008

Ferguson, C. (2009). An effect size primer: A guide for clinicians and researchers. *Professional Psychology: Research and Practice, 40*(5), 532–538. DOI:10.1037/a0015808

Ferguson, E. (2013). Personality is of central concern to understand health: Towards a theoretical model for health psychology. *Health Psychology Review, 7*(suppl. 1), S32–S70. DOI:10.1080/17437199.2010.547985

Ferreira, M., Fischer, R., Porto, J., Pilati, R., & Milfont, T. (2011). Unraveling the mystery of Brazilian jeitinho: A cultural exploration of social norms. *Personality & Social Psychology Bulletin, 38*(3), 331–344. DOI:10.1177/0146167211427148

Ferreira, M., O'Donovan, M., Meng, Y., Jones, I., Ruderfer, D., Jones, L., . . . Craddock, N. (2008). Collaborative genome-wide association analysis supports a role for ANK3 and CACNA1C in bipolar disorder. *Nature Genetics, 40*(9), 1056–1058. DOI:10.1038/ng.209

Fincher, C., & Thornhill, R. (2012). Parasite-stress promotes in-group assortative sociality: The cases of strong family ties and heightened religiosity. *Behavioral and Brain Sciences, 35*(2), 61–79. DOI:10.1017/s0140525x11000021

Fincher, C., Thornhill, R., Murray, D., & Schaller, M. (2008). Pathogen prevalence predicts human cross-cultural variability in individualism/collectivism. *Proceedings of the Royal Society B: Biological Sciences, 275*(1640), 1279–1285. DOI:10.1098/rspb.2008.0094

Fink, B., Neave, N., Manning, J., & Grammer, K. (2005). Facial symmetry and the "big-five" personality factors. *Personality and Individual Differences, 39*(3), 523–529. DOI:10.1016/j.paid.2005.02.002

Fink, B., Weege, B., Flügge, J., Röder, S., Neave, N., & McCarty, K. (2012). Men's personality and women's perception of their dance quality. *Personality and Individual Differences, 52*(2), 232–235. DOI:10.1016/j.paid.2011.10.008

Fischer, R. (2013a). Gene-environment interactions are associated with endorsement of social hierarchy values and beliefs across cultures. *Journal of Cross-Cultural Psychology, 44*, 1107–1121. DOI:10.1177/0022022112471896

Fischer, R. (2013b). What values can (and cannot) tell us about individuals, society and culture. *Advances in Culture and Psychology, 4*, 218–272.

Fischer, R., & Boer, D. (2011). What is more important for national well-being: Money or autonomy? A meta-analysis of well-being, burnout, and anxiety across 63 societies. *Journal of Personality and Social Psychology, 101*(1), 164–184. DOI:10.1037/a0023663

Fischer, R., & Boer, D. (2015). Motivational basis of personality traits: A meta-analysis of value-personality correlations. *Journal of Personality, 83*(5), 491–510. DOI:10.1111/jopy.12125

Fischer, R., & Fontaine, J. (2011). Methods for investigating structural equivalence. In D. Matsumoto & F. van de Vijver (Eds.), *Cross-cultural research methods in psychology* (pp. 179–215). Cambridge: Cambridge University Press.

Fischer, R., Fontaine, J., van de Vijver, F., & van Hemert, D. (2009). An examination of acquiescent response styles in cross-cultural research. In A. Gari & K. Mylonas (Eds.), *Quod Erat Demonstrandum: From Herodotus' ethnographic journeys to cross-cultural research* (pp. 137–148). Pedio: Athens.

Fischer, R., Lee, A., & Verzeijden, M. N. (2017). *Thermal stress can bring out the best: Dopamine genes are linked to personality traits, but only in demanding climates.* Victoria: University of Wellington.

Fischer, R., Milfont, T., & Gouveia, V. (2011). Does social context affect value structures? Testing the within-country stability of value structures with a functional theory of values. *Journal of Cross-Cultural Psychology, 42*(2), 253–270. DOI:10.1177/0022022110396888

Fischer, R., & Schwartz, S. H. (2010). Whence differences in value priorities? Individual, cultural, or artifactual sources. *Journal of Cross-Cultural Psychology, 42*(7), 1127–1144. DOI:10.1177/0022022110381429

Fischer, R., Vauclair, C.-M., Fontaine, J., & Schwartz, S. (2010). Are individual-level and country-level value structures different? Testing Hofstede's legacy with the Schwartz Value Survey. *Journal of Cross-Cultural Psychology, 41*, 135–151.

Fleeson, W. (2001). Toward a structure- and process-integrated view of personality: Traits as density distributions of states. *Journal of Personality and Social Psychology, 80*(6), 1011–1027. DOI:10.1037//0022-3514.80.6.1011

Fleeson, W., & Jayawickreme, E. (2015). Whole trait theory. *Journal of Research in Personality, 56*, 82–92. DOI:10.1016/j.jrp.2014.10.009

Flynn, E., Laland, K., Kendal, R., & Kendal, J. (2013). Developmental niche construction. *Developmental Science, 16*(2), 296–313. DOI:10.1111/desc.12030

Fontaine, J. R. (2005). Equivalence. *Encyclopedia of Social Measurement* (vol. 1, pp. 803–813). New York: Academic Press.

Fontaine, J. R., Poortinga, Y. H., Delbeke, L., & Schwartz, S. H. (2008). Structural equivalence of the values domain across cultures: Distinguishing sampling fluctuations from meaningful variation. *Journal of Cross-Cultural Psychology, 39*(4), 345–365. DOI:10.1177/0022022108318112

Fontaine, J. R., Scherer, K., Roesch, E., & Ellsworth, P. (2007). The world of emotions is not two-dimensional. *Psychological Science, 18*(12), 1050–1057. DOI:10.1111/j.1467-9280.2007.02024.x

Foster, J., & Neufeld, K.-A. (2013). Gut–brain axis: How the microbiome influences anxiety and depression. *Trends in Neurosciences, 36*(5), 305–312. DOI:10.1016/j.tins.2013.01.005

Fouts, H. N., Hewlett, B. S., & Lamb, M. E. (2012). A biocultural approach to breast-feeding interactions in Central Africa. *American Anthropologist*, *114*(1), 123–136. DOI:10.1111/j.1548–1433.2011.01401.x

Freeman, H. D., & Gosling, S. D. (2010). Personality in nonhuman primates: Are-view and evaluation of past research. *American Journal of Primatology*, *71*, 1–19. DOI:10.1002/ajp.20833

Frese, M., Kring, W., Soose, A., & Zempel, J. (1996). Personal initiative at work: Differences between East and West Germany. *Academy of Management Journal*, *39*, 37–63.

Galton, F. (1949). The measurement of character. *Fortnightly Review*, *36*, 179–185.

Goldberg, L. (1981). Language and individual differences: The search for universals in personality lexicons. *Review of Personality and Social Psychology*, 2, 141–165.

Goldberg, L. (1990). An alternative "description of personality": The Big-Five factor structure. *Journal of Personality and Social Psychology*, *59*(6), 1216–1229. DOI:10.1037/0022–3514.59.6.1216

Goldberg, L. (1993). The structure of phenotypic personality traits. *American Psychologist*, *48*(1), 26–34. DOI:10.1037/0003–066x.48.1.26

Goldberg, L., Johnson, J., Eber, H., Hogan, R., Ashton, M., Cloninger, R., & Gough, H. (2006). The International Personality Item Pool and the future of public-domain personality measures. *Journal of Research in Personality*, *40*(1), 84–96. DOI:10.1016/j.jrp.2005.08.007

Gosling, S. (2001). From mice to men: What can we learn about personality from animal research? *Psychological Bulletin*, 127, 45–86.

Gosling, S. (2008). Personality in non-human animals. *Social and Personality Psychology Compass*, 2(2), 985–1001. DOI:10.1111/j.1751–9004.2008.00087.x

Gouveia, V., Milfont, T., & Guerra, V. (2014). Functional theory of human values: Testing its content and structure hypotheses. *Personality and Individual Differences*, *60*, 41–47.

Gouveia, V., Vione, K., Milfont, T., & Fischer, R. (2015). Patterns of value change during the life span. *Personality and Social Psychology Bulletin*, *41*(9), 1276–1290. DOI:10.1177/0146167215594189

Grant, P. R., & Grant, R. B. (2006). Evolution of character displacement in Darwin's finches. *Science*, *313*(5784), 224–226. DOI:10.1126/science.1128374

Gray, J. (1970). The psychophysiological basis of introversion-extraversion. *Behaviour Research and Therapy*, *8*(3), 249–266.

Gray, J. (1975). *Elements of a two-process theory of learning*. London: Academic Press.

Grouzet, F., Kasser, T., Ahuvia, A., Dols, J., Kim, Y., Lau, S., . . . Sheldon, K. (2005). The structure of goal contents across 15 cultures. *Journal of Personality and Social Psychology*, *89*(5), 800–816.

Guilford, J. P. (1959). *Personality*. New York: McGraw-Hill.

Gurven, M., von Rueden, C., Massenkoff, M., Kaplan, H., & Vie, M. (2013). How universal is the Big Five? Testing the five-factor model of personality variation among forager–farmers in the Bolivian Amazon. *Journal of Personality and Social Psychology*, *104*(2), 354–370. DOI:10.1037/a0030841

Hamilton, W. D., & Zuk, M. (1982). Heritable true fitness and bright birds: A role for parasites? *Science*, *218*(4570), 384–387.

Hancock, A., Witonsky, D., Ehler, E., Alkorta-Aranburu, G., Beall, C., Gebremedhin, A., . . . Rienzo, A. (2010). Human adaptations to diet, subsistence, and ecoregion are due to subtle shifts in allele frequency. *Proceedings of the National Academy of Sciences, 107* (suppl. 2), 8924–8930. DOI:10.1073/pnas.0914625107

Hariri, A. (2009). The neurobiology of individual differences in complex behavioral traits. *Annual Review of Neuroscience, 32*(1), 225–247. DOI:10.1146/annurev.neuro.051508.135335

Heider, F. (1946). Attitudes and cognitive organization. *Journal of Psychology, 21*(1), 107–112. DOI:10.1080/00223980.1946.9917275

Heijmans, B., Tobi, E., Stein, A., Putter, H., Blauw, G., Susser, E., . . . Lumey. (2008). Persistent epigenetic differences associated with prenatal exposure to famine in humans. *Proceedings of the National Academy of Sciences, 105*(44), 17046–17049. DOI:10.1073/pnas.0806560105

Hemert, D. A., de Vijver, F., Poortinga, Y. H., & Georgas, J. (2002). Structural and functional equivalence of the Eysenck Personality Questionnaire within and between countries. *Personality and Individual Differences, 33*, 1229–1249.

Hewlett, B., & Hewlett, B. S. (2012). Hunter-gatherer adolescence. In B. L. Hewlett (Ed.), *Adolescent identity: Evolutionary, cultural and developmental perspectives* (pp. 73–101). London: Routledge.

Hewlett, B. S., & Lamb, M. E. (2005). *Hunter-gatherer childhoods: Evolutionary, developmental, and cultural perspectives*. New Brunswick, NJ: Transaction.

Hewlett, B. S., & Roulette, C. J. (2016). Teaching in hunter-gatherer infancy. *Royal Society Open Science, 3*, 150403. oi.org/10.1098/rsos.150403

Hieronymus, F., Emilsson, J. F., Nilsson, S., & Eriksson, E. (2016). Consistent superiority of selective serotonin reuptake inhibitors over placebo in reducing depressed mood in patients with major depression. *Molecular Psychiatry, 21*(4), 523–530. DOI:10.1038/mp.2015.53

Higgins, E. T. (1998). Promotion and prevention: Regulatory focus as a motivational principle. In M. P. Zanna (Ed.), *Advances in experimental social psychology* (vol. 30, pp. 1–46). New York: Academic Press.

Hill, C., Nel, J., de Vijver, F., Meiring, D., Valchev, V., Adams, B., & Bruin, G. (2013). Developing and testing items for the South African Personality Inventory (SAPI). *SA Journal of Industrial Psychology, 39*(1), 1122. DOI:10.4102/sajip.v39i1.1122

Hofstede, G. (1980). *Culture's consequences: National differences in thinking and organizing*. Beverly Hills: Sage.

Hofstede, G., & McRae, R. (2004). Personality and culture revisited: Linking traits and dimensions of culture. *Cross-Cultural Research, 38*(1), 52–88. DOI:10.1177/1069397103259443

Holland, J. (1996). Exploring careers with a typology: What we have learned and some new directions. *American Psychologist, 51*(4), 397. DOI:10.1037/0003-066x.51.4.397

Holzinger, K., & Swineford, F. (1937). The bi-factor method. *Psychometrika, 2*, 41–54.

Inglehart, R. (1971). The silent revolution in Europe: Intergenerational change in post-industrial societies. *American Political Science Review, 65*, 991–1017.

Inglehart, R. (1990). *Culture shift in advanced industrial society*. Princeton: Princeton University Press.

Inglehart, R. (1997). *Modernization and postmodernization: Cultural, economic and political changes in 43 societies*.Princeton: Princeton University Press.

Inglehart, R., & Baker, W. (2000). Modernization, cultural change, and the persistence of traditional values. *American Sociological Review, 65*, 19–51.

Inglehart, R., & Welzel, C. (2005). *Modernization, cultural change, and democracy: The human development sequence.* Cambridge: Cambridge University Press.

Ingram, C. J., Liebert, A., & Swallow, D. M. (2012). *Population Genetics of Lactase Persistence and Lactose Intolerance.* In eLS: Evolution & Diversity of life. Chichester: John Wiley & Sons. DOI:10.1002/9780470015902.a0020855.pub2

Jaffee, S., McFarquhar, T., Stevens, S., Ouellet-Morin, I., Melhuish, E., & Belsky, J. (2015). Interactive effects of early and recent exposure to stressful contexts on cortisol reactivity in middle childhood. *Journal of Child Psychology and Psychiatry, 56*(2), 138–146. DOI:10.1111/jcpp.12287

John, O., Naumann, L., & Soto, C. (2008). Paradigm shift to the integrative big five trait taxonomy: History, measurement and conceptual issues. In O. P. John, R. W. Robins, & L. A. Pervin (Eds.), *Handbook of personality: Theory and research* (3rd ed, pp. 114–158). New York: Guilford Press.

John-Henderson, N., Stellar, J., Mendoza-Denton, R., & Francis, D. (2015). Socioeconomic status and social support. *Psychological Science, 26*(10), 1620–1629. DOI:10.1177/0956797615595962

Johnson, W., Penke, L., & Spinath, F. (2011). Heritability in the era of molecular genetics: Some thoughts for understanding genetic influences on behavioural traits. *European Journal of Personality, 25*(4), 254–266. DOI:10.1002/per.836

Karlan, & Zinman. (2011). Microcredit in theory and practice: Using randomized credit scoring for impact evaluation. *Science, 332*(6035), 1278–1284. DOI:10.1126/science.1200138

Keller, L., Bouchard, T., Arvey, R., Segal, N., & Dawis, R. (1992). Work values: Genetic and environmental influences. *Journal of Applied Psychology, 77*(1), 79–88. DOI:10.1037/0021-9010.77.1.79

Keltner, D., Kogan, A., Piff, P., & Saturn, S. (2014). The sociocultural appraisals, values, and emotions (SAVE) framework of prosociality: Core processes from gene to meme. *Annual Review of Psychology, 65*(1), 425–460. DOI:10.1146/annurev-psych-010213-115054

Kenrick, Griskevicius, Neuberg, & Schaller. (2010). Renovating the pyramid of needs: Contemporary extensions built upon ancient foundations. *Perspectives on Psychological Science, 5*(3), 292–314. DOI:10.1177/1745691610369469

Kidd, C., & Hayden, B. Y. (2015). The psychology and neuroscience of curiosity. *Neuron, 88*(3), 449–460. DOI:10.1016/j.neuron.2015.09.010

Kitayama, S., & Uskul, A. (2011). Culture, mind, and the brain: Current evidence and future directions. *Annual Review of Psychology, 62*(1), 419–449. DOI:10.1146/annurev-psych-120709-145357

Kluckhohn, F. R., & Strodtbeck, F. L. (1961). *Variations in value orientations.* Evanston, IL: Row, Peterson.

Knafo, A., & Spinath, F. (2011). Genetic and environmental influences on girls' and boys' gender-typed and gender-neutral values. *Developmental Psychology, 47*(3), 726–731. DOI:10.1037/a0021910

Konner, M. (2005). Hunter-gatherer infancy and childhood. In B. S. Hewlett & M. E. Lamb (Eds.), *Hunter-gatherer childhoods: Evolutionary, developmental, and cultural perspectives* (pp. 19–64). New Brunswick, NJ: Transaction.

Konner, M. (2016). Hunter-gatherer infancy and childhood in the context of human evolution. In C. L. Meehan & A. N. Crittenden (Eds.), *Childhood: Origins, evolution, and implications* (pp. 123–154). Santa Fe: School of Advanced Research Press.

Kraus, M., & Callaghan, B. (2016). Social class and prosocial behavior. *Social Psychological and Personality Science, 7*(8), 769–777. DOI:10.1177/1948550616659120

Kraus, M., Piff, P. K., & Keltner, D. (2009). Social class, sense of control, and social explanation. *Journal of Personality and Social Psychology, 97*(6), 992–1004. DOI:10.1037/a0016357

Kraus, M., Piff, P., Mendoza-Denton, R., Rheinschmidt, M., & Keltner, D. (2012). Social class, solipsism, and contextualism: How the rich are different from the poor. *Psychological Review, 119*(3), 546–572. DOI:10.1037/a0028756

Kruger, AC, & Konner, M. (2010). Who responds to crying? *Human Nature, 21*, 309–329.

Laayouni, H., Oosting, M., Luisi, P., Ioana, M., Alonso, S., Ricaño-Ponce, I., . . . Netea, M. (2014). Convergent evolution in European and Roma populations reveals pressure exerted by plague on Toll-like receptors. *Proceedings of the National Academy of Sciences, 111*(7), 2668–2673. DOI:10.1073/pnas.1317723111

Laland, K. N., & Brown, G. R. (2011). *Sense and nonsense: Evolutionary perspectives on human behaviour* (2nd ed.). Oxford: Oxford University Press.

Laland, K., Odling-Smee, J., & Feldman, M. (2000). Niche construction, biological evolution, and cultural change. *Behavioral and Brain Sciences, 23*(1), 131–146. DOI:10.1017/s0140525x00002417

Lane, A., Luminet, O., Nave, G., & Mikolajczak, M. (2016). Is there a publication bias in behavioural intranasal oxytocin research on humans? Opening the file drawer of one laboratory. *Journal of Neuroendocrinology, 28*(4), 1–6. DOI:10.1111/jne.12384

Latzman, R., Freeman, H., Schapiro, S., & Hopkins, W. (2015). The contribution of genetics and early rearing experiences to hierarchical personality dimensions in chimpanzees (Pan troglodytes). *Journal of Personality and Social Psychology, 109*, 889–900.

Latzman, R., Sauvigné, K., & Hopkins, W. (2016). Translating chimpanzee personality to humans: Investigating the transportability of chimpanzee-derived personality scales to humans. *American Journal of Primatology, 78*(6), 601–609. DOI:10.1002/ajp.22522

Lazarus, R., & Folkman, S. (1984). *Stress, appraisal and coping.* New York: Springer.

Lee, K., Ashton, M., Pozzebon, J., Visser, B., Bourdage, J., & Ogunfowora, B. (2009). Similarity and assumed similarity in personality reports of well-acquainted persons. *Journal of Personality and Social Psychology, 96*(2), 460. DOI:10.1037/a0014059

Leonardi, M., Librado, P., Sarkissian, C., Schubert, M., Alfarhan, A., Alquraishi, S., . . . Orlando, L. (2016). Evolutionary patterns and processes: Lessons from ancient DNA. *Systematic Biology, 66*, e1–e29.

Lesch, Bengel, Heils, Sabol, Greenberg, Petri, . . . Murphy. (1996). Association of anxiety-related traits with a polymorphism in the serotonin transporter gene regulatory region. *Science*, *274*(5292), 1527–1531.

Lewis, D. (2015). Evolved individual differences: Advancing a condition-dependent model of personality. *Personality and Individual Differences*, *84*, 63–72. DOI:10.1016/j.paid.2014.10.013

Lindeman, M., & Verkasalo, M. (2005). Measuring values with the Short Schwartz's Value Survey. *Journal of Personality Assessment*, *85*(2), 170–178. DOI:10.1207/s15327752jpa8502_09

Litman, J. (2005). Curiosity and the pleasures of learning: Wanting and liking new information. *Cognition & Emotion*, *19*(6), 793–814. DOI:10.1080/02699930541000101

Llamas, B., Willerslev, E., & Orlando, L. (2017). Human evolution: A tale from ancient genomes. *Philosophical Transactions of the Royal Society of London. Series B, Biological Sciences*, *372*(1713).

Lo, M.-T., Hinds, D., Tung, J., Franz, C., Fan, C.-C., Wang, Y., . . . Chen, C.-H. (2016). Genome-wide analyses for personality traits identify six genomic loci and show correlations with psychiatric disorders. *Nature Genetics*, *49*, 152–156. DOI:10.1038/ng.3736

Lodi-Smith, & Roberts, B. (2007). Social investment and personality: A meta-analysis of the relationship of personality traits to investment in work, family, religion, and volunteerism. *Personality and Social Psychology Review*, *11*(1), 68–86. DOI:10.1177/1088868306294590

Lüdtke, O., Roberts, B., Trautwein, U., & Nagy, G. (2011). A random walk down university avenue: Life paths, life events, and personality trait change at the transition to university life. *Journal of Personality and Social Psychology*, *101*(3), 620–637. DOI:10.1037/a0023743

Lukaszewski, A. W. & Roney, J. R. (2011). The origins of extraversion: Joint effects of facultative calibration and genetic polymorphism. *Personality and Social Psychology Bulletin*, *37*(3), 409–421. DOI:10.1177/0146167210397209

Lun, V., Fischer, R., & Ward, C. (2010). Exploring cultural differences in critical thinking: Is it about my thinking style or the language I speak? *Learning and Individual Differences*, *20*(6), 604–616. DOI:10.1016/j.lindif.2010.07.001

MacDonald, K. (1995). Evolution, the Five-Factor Model, and levels of personality. *Journal of Personality*, *63*(3), 525–567. DOI:10.1111/j.1467–6494.1995.tb00505.x

Maio, G. (2010). Mental representations of social values. *Advances in Experimental Social Psychology*, *42*, 1–43.

Maio, G. R. (2016). *The psychology of human values*. London: Psychology Press.

Maio, G., & Olson, J. (1998). Values as truisms: Evidence and implications. *Journal of Personality and Social Psychology*, *74*(2), 294. DOI:10.1037/0022–3514.74.2.294

Maraun, M. (1997). Appearance and reality: Is the Big Five the structure of trait descriptors? *Personality and Individual Differences*, *22*(5), 629–647. DOI:10.1016/s0191–8869(96)00262–0

Marcia, J. E. (1966). Development and validation of ego-identity status. *Journal of Personality and Social Psychology*, *3*(5), 551–558.

Markey, P., Markey, C., & Tinsley, B. (2004). Children's behavioral manifestations of the Five-Factor Model of personality. *Personality and Social Psychology Bulletin, 30*(4), 423–432. DOI:10.1177/0146167203261886

Markey, P., Markey, C., Ericksen, A. J., & Tinsley, B. J. (2002). A preliminary validation of preadolescents' self-reports using the Five-Factor Model of personality. *Journal of Research in Personality*, 36(2), 173–181.

Maslow, A. (1943). A theory of human motivation. *Psychological Review, 50*, 370–396.

Maynard Smith, J., & Price, G. R. (1973). The logic of animal conflict. *Nature, 246*(5427), 15–18. DOI:10.1038/246015a0

McAdams, D. (1995). What do we know when we know a person? *Journal of Personality, 63*(3), 365–396. DOI:10.1111/j.1467-6494.1995.tb00500.x

McAdams, D., & Pals, J. (2006). A new Big Five: Fundamental principles for an integrative science of personality. *American Psychologist, 61*(3), 204–217. DOI:10.1037/0003-066X.61.3.204

McCrae, R. (2001). Trait psychology and culture: Exploring intercultural comparisons. *Journal of Personality, 69*(6), 819–846. DOI:10.1111/1467-6494.696166

McCrae, R. R., & Costa, P. T. (1999). A five-factor theory of personality. *Handbook of Personality: Theory and Research*, 2, 159–181.

McCrae, R. R., Jang, K. L., Livesley, W. J., Riemann, R., & Angleitner, A. (2001). Sources of structure: Genetic, environmental, and artifactual influences on the covariation of personality traits. *Journal of Personality, 69*(4), 511–535. DOI:10.1111/1467-6494.694154

McCrae, R., & John, O. (1992). An introduction to the Five-Factor Model and its applications. *Journal of Personality, 60*(2), 175–215. DOI:10.1111/j.1467-6494.1992.tb00970.x

McCrae, R., Jr., P., Ostendorf, F., Angleitner, A., Hřebíčková, M., Avia, M., . . . Smith, P. (2000). Nature over nurture: Temperament, personality, and life span development. *Journal of Personality and Social Psychology, 78*(1), 173–186. DOI:10.1037/0022-3514.78.1.173

McCrae, R. R., Terracciano, A., & Personality Profiles of Cultures Project, M. (2005). Universal features of personality traits from the observer's perspective: Data from 50 cultures. *Journal of Personality and Social Psychology, 88*(3), 547–561. DOI:10.1037/0022-3514.88.3.547

Meredith, W. (1993). Measurement invariance, factor analysis and factorial invariance. *Psychometrika, 58*(4), 525–543. DOI:10.1007/bf02294825

Mesoudi, A. (2009). How cultural evolutionary theory can inform social psychology and vice versa. *Psychological Review, 116*(4), 929–952. DOI:10.1037/a0017062

Mesoudi, A., Whiten, A., & Laland, K. (2006). Towards a unified science of cultural evolution. *Behavioral and Brain Sciences*, 29(04), 329–347; discussion 347–383. DOI:10.1017/s0140525x06009083

Milfont, T., Milojev, P., & Sibley, C. (2016). Values stability and change in adulthood: A 3-year longitudinal study of rank-order stability and mean-level differences. *Personality & Social Psychology Bulletin*, 42(5), 572–588. DOI:10.1177/0146167216639245

Milojev, P., Osborne, D., Greaves, L., Barlow, F., & Sibley, C. (2013). The Mini-IPIP6: Tiny yet highly stable markers of Big Six personality. *Journal of Research in Personality, 47*(6), 936–944. DOI:10.1016/j.jrp.2013.09.004

Mischel. (1973). Toward a cognitive social learning reconceptualization of personality. *Psychological Review, 80*(4), 252–283.

Mischel, W. (1977). The interaction of person and situation. In D. Magnusson & N. S. Endler (Eds.), *Personality at the crossroads: Current issues in interactional psychology* (pp. 333–352). Hillsdale, NJ: Erlbaum.

Mischel, W., & Shoda, Y. (1995). A cognitive-affective system theory of personality: Reconceptualizing situations, dispositions, dynamics, and invariance in personality structure. *Psychological Review, 102*(2), 246–268. DOI:10.1037//0033-295x.102.2.246

Moffitt, T., Arseneault, L., Belsky, D., Dickson, N., Hancox, R., Harrington, H., . . . Caspi, A. (2011). A gradient of childhood self-control predicts health, wealth, and public safety. *Proceedings of the National Academy of Sciences, 108*(7), 2693–2698. DOI:10.1073/pnas.1010076108

Mujcic, R., & Oswald, A. J. (2016). Evolution of well-being and happiness after increases in consumption of fruit and vegetables. *American Journal of Public Health, 106*(8), 1504–1510. DOI:10.2105/ajph.2016.303260

Munafò, M., & Flint, J. (2011). Dissecting the genetic architecture of human personality. *Trends in Cognitive Sciences, 15*(9), 395–400. DOI:10.1016/j.tics.2011.07.007

Munafò, M., Yalcin, B., Willis-Owen, S., & Flint, J. (2007). Association of the dopamine D4 receptor (DRD4) gene and approach-related personality traits: Meta-analysis and new data. *Biological Psychiatry, 63*(2), 197–206. DOI:10.1016/j.biopsych.2007.04.006

Murray, D., & Schaller, M. (2009). Historical prevalence of infectious diseases within 230 geopolitical regions: A tool for investigating origins of culture. *Journal of Cross-Cultural Psychology, 41*(1), 99–108. DOI:10.1177/0022022109349510

Murray, D., Trudeau, R., & Schaller, M. (2011). On the origins of cultural differences in conformity: Four tests of the pathogen prevalence hypothesis. *Personality and Social Psychology Bulletin, 37*(3), 318–329. DOI:10.1177/0146167210394451

Nave, G., Camerer, C., & McCullough, M. (2015). Does oxytocin increase trust in humans? A critical review of research. *Perspectives on Psychological Science, 10*(6), 772–789. DOI:10.1177/1745691615600138

Nel, J., Valchev, V., Rothmann, S., Vijver, F., Meiring, D., & Bruin, G. (2012). Exploring the personality structure in the 11 languages of South Africa. *Journal of Personality, 80*(4), 915–948. DOI:10.1111/j.1467-6494.2011.00751.x

Nettle, D. (2005). An evolutionary approach to the extraversion continuum. *Evolution and Human Behavior, 26*(4), 363–373. DOI:10.1016/j.evolhumbehav.2004.12.004

Nettle, D. (2006). The evolution of personality variation in humans and other animals. *American Psychologist, 61*(6), 622. DOI:10.1037/0003-066x.61.6.622

Nettle, D., Buss, D. M., & Hawley. (2011). *Evolutionary perspectives on the five-factor model of personality*. In D. M. Buss & P. H. Hawley (Eds.), *The evolution of personality and individual differences* (pp. 5–28). Oxford: Oxford University Press.

Nettle, D., Frankenhuis, W., & Rickard, I. (2013). The evolution of predictive adaptive responses in human life history. *Proceedings of the Royal Society of London B: Biological Sciences, 280*(1766). DOI:10.1098/rspb.2013.1343

Niehoff, E., Petersdotter, L., & Freund, P. (2017). International sojourn experience and personality development: Selection and socialization effects of studying abroad and the Big Five. *Personality and Individual Differences, 112*, 55–61.

Nilsson, A. (2014). Personality psychology as the integrative study of traits and worldviews. *New Ideas in Psychology*, 32, 18–32. DOI:10.1016/j.newideapsych.2013.04.008

Pace, V., & Brannick, M. (2010). How similar are personality scales of the "same" construct? A meta-analytic investigation. *Personality and Individual Differences, 49*(7), 669–676. DOI:10.1016/j.paid.2010.06.014

Parks-Leduc, L., Feldman, G., & Bardi, A. (2014). Personality traits and personal values: A meta-analysis. *Personality and Social Psychology Review, 19*(1), 3–29. DOI:10.1177/1088868314538548

Penke, L., Denissen, J., & Miller, G. (2007). The evolutionary genetics of personality. *European Journal of Personality, 21*(5), 549–587. DOI:10.1002/per.629

Piff, P., Kraus, M., Côté, S., Cheng, B., & Keltner, D. (2010). Having less, giving more: The influence of social class on prosocial behavior. *Journal of Personality and Social Psychology, 99*, 771–784.

Piff, P., Stancato, D., Côté, S., Mendoza-Denton, R., & Keltner, D. (2012). Higher social class predicts increased unethical behavior. *Proceedings of the National Academy of Sciences, 109*(11), 4086–4091. DOI:10.1073/pnas.1118373109

Piff, P., Stancato, D., Martinez, A., Kraus, M., & Keltner, D. (2012). Class, chaos, and the construction of community. *Journal of Personality and Social Psychology, 103*(6), 949–962. DOI:10.1037/a0029673

Polderman, T., Benyamin, B., Leeuw, C., Sullivan, P., Bochoven, A., Visscher, P., & Posthuma, D. (2015). Meta-analysis of the heritability of human traits based on fifty years of twin studies. *Nature Genetics, 47*(7), 702–709. DOI:10.1038/ng.3285

Poortinga, Y. H. (2015). Is "culture" a workable concept for (cross-) cultural psychology? *Online Readings in Psychology and Culture*, 2(1). https://doi.org/10.9707/2307-0919.1139

Poortinga, Y. H., & van de Vijver, F. J. R. (1987). Explaining cross-cultural differences: Bias analysis and beyond. *Journal of Cross-Cultural Psychology, 18*, 259–282.

Pound, N., Penton-Voak, I., & Brown, W. (2007). Facial symmetry is positively associated with self-reported extraversion. *Personality and Individual Differences, 43*(6), 1572–1582. DOI:10.1016/j.paid.2007.04.014

Rammstedt, B., & John, O. P. (2007). Measuring personality in one minute or less: A 10-item short version of the Big Five Inventory in English and German. *Journal of Research in Personality, 41*, 203–212.

Ratner, C. (2009). The cultural psychology of oppression and liberation. *Journal of Social Distress and the Homeless*, 18, 231–268. DOI:10.1179/105307809805365190

Read, S., Monroe, B., Brownstein, A., Yang, Y., Chopra, G., & Miller, L. (2010). A neural network model of the structure and dynamics of human personality. *Psychological Review, 117*(1), 61–92. DOI:10.1037/a0018131

Read, S., & Miller, L. C. (2002). Virtual personalities: A neural network model of personality. *Personality and Social Psychology Review, 6*, 357–369.

Reist, C., Ozdemir, V., Wang, E., Hashemzadeh, M., Mee, S., & Moyzis, R. (2007). Novelty seeking and the dopamine D4 receptor gene (DRD4) revisited in

Asians: Haplotype characterization and relevance of the 2-repeat allele. *American Journal of Medical Genetics. Part B, Neuropsychiatric Genetics, 144B*(4), 453–457. DOI:10.1002/ajmg.b.30473

Renner, W. (2003). Human values: A lexical perspective. *Personality and Individual Differences, 34*(1), 127–141. DOI:10.1016/s0191–8869(02)00037–5

Renner, W., Kandler, C., Bleidorn, W., Riemann, R., Angleitner, A., Spinath, F., & Menschik-Bendele, J. (2012). Human values: Genetic and environmental effects on five lexically derived domains and their facets. *Personality and Individual Differences, 52*(1), 89–93. DOI:10.1016/j.paid.2011.09.003

Richerson, P. J., & Boyd, R. (2005). *Not by genes alone: How culture transformed human evolution.* Chicago: Chicago University Press.

Roberts, B. (1997). Plaster or plasticity: Are adult work experiences associated with personality change in women? *Journal of Personality, 65*(2), 205–232. DOI:10.1111/j.1467–6494.1997.tb00953.x

Roberts, B., & Davis, J. P. (2016). Young adulthood is the crucible of personality development. *Emerging Adulthood*, 1–9. DOI:10.1177/2167696816653052

Roberts, B., Luo, J., Briley, D., Chow, P., Su, R., & Hill, P. (2017). A systematic review of personality trait change through intervention. *Psychological Bulletin, 143*(2), 117–141. DOI:10.1037/bul0000088

Roberts, B., Walton, K., & Viechtbauer, W. (2006). Patterns of mean-level change in personality traits across the life course: A meta-analysis of longitudinal studies. *Psychological Bulletin, 132*(1), 1-25. DOI:10.1037/0033–2909.132.1.1

Roberts, B., Wood, D., & Smith, J. (2005). Evaluating Five Factor Theory and social investment perspectives on personality trait development. *Journal of Research in Personality, 39*(1), 166–184. DOI:10.1016/j.jrp.2004.08.002

Roccas, S., Sagiv, L., Schwartz, S. H., & Knafo, A. (2002). The Big Five personality factors and personal values. *Personality and Social Psychology Bulletin, 28*(6), 789–801. DOI:10.1177/0146167202289008

Rodriguez, A., Reise, S. P., & Haviland, M. G. (2016). Applying bifactor statistical indices in the evaluation of psychological measures. *Journal of Personality Assessment, 98*, 223–237.

Rokeach, M. (1968). *Beliefs, attitudes and values: A theory of organization and change.* San Francisco: Jossey-Bass.

Rokeach, M. (1973). *The nature of human values.* New York: Free Press.

Rosenström, T., Jylhä, P., Pulkki-Råback, L., Holma, M., Raitakari, O., Isometsä, E., & Keltikangas-Järvinen, L. (2015). Long-term personality changes and predictive adaptive responses after depressive episodes. *Evolution and Human Behavior, 36*(5), 337–344. DOI:10.1016/j.evolhumbehav.2015.01.005

Salgado, J. F. (2003). Predicting job performance using FFM and non-FFM personality measures. *Journal of Occupational and Organizational Psychology, 76*, 323–346.

Saucier, G., Kenner, J., Iurino, K., Malham, P., Chen, Z., Thalmayer, A., . . . Altschul, C. (2014). Cross-cultural differences in a global "survey of world views." *Journal of Cross-Cultural Psychology, 46*(1), 53–70. DOI:10.1177/0022022114551791

Saucier, G., Thalmayer, A. G., & Bel-Bahar, T. S. (2014). Human attribute concepts: Relative ubiquity across twelve mutually isolated languages. *Journal of Personality and Social Psychology, 107*(1), 199–216. DOI:10.1037/a0036492

Saucier, G., Thalmayer, A., Payne, D., Carlson, R., Sanogo, L., Ole-Kotikash, L., . . . Zhou, X. (2014). A basic bivariate structure of personality attributes evident across nine languages. *Journal of Personality, 82*(1), 1–14. DOI:10.1111/jopy.12028

Schaller, M., & Murray, D. (2008). Pathogens, personality, and culture: Disease prevalence predicts worldwide variability in sociosexuality, extraversion, and openness to experience. *Journal of Personality and Social Psychology, 95*(1), 212–221. DOI:10.1037/0022–3514.95.1.212

Schaller, M., & Murray, D. (2011). Infectious disease and the creation of culture. *Advances in Culture and Psychology, 1*, 99–151.

Schein, E. (1985). *Organizational culture and leadership: A dynamic view*. San Francisco: Jossey-Bass.

Schermer, J., Feather, Zhu, G., & Martin, N. (2008). Phenotypic, genetic, and environmental properties of the portrait values questionnaire. *Twin Research and Human Genetics, 11*(5), 531–537. DOI:10.1375/twin.11.5.531

Schermer, J., Vernon, P., Maio, G., & Jang, K. (2011). A behavior genetic study of the connection between social values and personality. *Twin Research and Human Genetics, 14*(3), 233–239. DOI:10.1375/twin.14.3.233

Schimmack, U., & Lucas, R. E. (2010). Environmental influences on well-being: A dyadic latent panel analysis of spousal similarity. *Social Indicators Research, 98*(1), 1–21. DOI: 10.1007/s11205-009-9516-8

Schmitt, D. P., Allik, J., McCrae, R. R., & Benet-Martinez, V. (2007). The geographic distribution of Big Five personality traits: Patterns and profiles of human self-description across 56 nations. *Journal of Cross-Cultural Psychology, 38*(2), 173–212. DOI:10.1177/0022022106297299

Schnorr, S., & Bachner, H. (2016). Integrative therapies in anxiety treatment with special emphasis on the gut microbiome. *Yale Journal of Biology and Medicine, 89*(3), 397–422.

Schwartz, S. H. (1992). Universals in the content and structure of values: Theoretical advances and empirical tests in 20 countries. *Advances in Experimental Social Psychology, 25*, 1–65.

Schwartz, S. H., & Bilsky, W. (1987). Toward a universal psychological structure of human values. *Journal of Personality and Social Psychology, 53*(3), 550–562. DOI:10.1037/0022–3514.53.3.550

Schwartz, S. H., & Bilsky, W. (1990). Toward a theory of the universal content and structure of values: Extensions and cross-cultural replications. *Journal of Personality and Social Psychology*. DOI: 10.1037/0022-3514.58.5.878

Schwartz, S. H., Cieciuch, J., Vecchione, M., Davidov, E., Fischer, R., Beierlein, C., . . . Konty, M. (2012). Refining the theory of basic individual values. *Journal of Personality and Social Psychology, 103*(4), 663–688. DOI:10.1037/a0029393

Schwartz, S. H., Melech, G., Lehmann, A., Burgess, S., Harris, M., & Owens, V. (2001). Extending the cross-cultural validity of the theory of basic human values with a different method of measurement. *Journal of Cross-Cultural Psychology, 32*, 519–542.

Sih, A., Bell, A., & Johnson, J. C. (2004). Behavioral syndromes: An ecological and evolutionary overview. *Trends in Ecology & Evolution, 19*(7), 372–378. DOI:10.1016/j.tree.2004.04.009

Singh, J. K., Misra, G., & De Raad, D. B. (2013). Personality structure in the trait lexicon of Hindi, a major language spoken in India. *European Journal of Personality, 27*, 605–620. DOI:10.1002/per.1940

Sireci, S. G. (2011). *Evaluating test and survey items for bias across languages and cultures*. In D. Matsumoto & F. van deVijver (Eds.), *Cross-cultural research methods in psychology* (pp. 216–240). Cambridge: Cambridge University Press.

Slobodskaya, H., & Kozlova, E. (2016). Early temperament as a predictor of later personality. *Personality and Individual Differences, 99*, 127–132. DOI:10.1016/j.paid.2016.04.094

Smith, P., & Fischer, R. (2008). Acquiescence, extreme response bias and culture: A multilevel analysis. In F. van deVijver, D. van Hemert, & Y. Poortinga (Eds.), *Individuals and cultures in multi-level analysis* (pp. 283–311). Mahwah, NJ: Erlbaum.

Smith, P. B., Fischer, R., Vignoles, V. L., & Bond, M. H. (2013). *Understanding social psychology across cultures: Engaging with others in a changing world* (2nd ed.). London: Sage.

Smith, P., Vignoles, V., Becker, M., Owe, E., Easterbrook, M., Brown, R., . . . Harb, C. (2016). Individual and culture-level components of survey response styles: A multi-level analysis using cultural models of selfhood. *International Journal of Psychology, 51*(6), 453–463. DOI:10.1002/ijop.12293

Soltis, J. (2004). The signal functions of early infant crying. *Behavioral and Brain Sciences, 27*(4), 443–458; discussion 459–490.

Sortheix, F., & Lönnqvist, J. (2014). Personal value priorities and life satisfaction in Europe: The moderating role of socioeconomic development. *Journal of Cross-Cultural Psychology, 45*, 282–299. DOI:10.1177/0022022113504621

Sortheix, F., & Lönnqvist, J. (2015). Person-group value congruence and subjective well-being in students from Argentina, Bulgaria and Finland: The role of interpersonal relationships. *Journal of Community & Applied Social Psychology, 25*(1), 34–48. DOI:10.1002/casp.2193

Sortheix, F., & Schwartz, S. (2017). Values that underlie and undermine well-being: variability across countries. *European Journal of Personality, 31*, 187–201. DOI:10.1002/per.2096

Sosis, R. (2000). Religion and intragroup cooperation: Preliminary results of a comparative analysis of utopian communities. *Cross-Cultural Research, 34*(1), 70–87. DOI:10.1177/106939710003400105

Sosis, R., & Bressler, E. (2003). Cooperation and commune longevity: A test of the costly signaling theory of religion. *Cross-Cultural Research, 37*(2), 211–239. DOI:10.1177/1069397103037002003

Soto, C., & John, O. (2009). Ten facet scales for the Big Five Inventory: Convergence with NEO PI-R facets, self-peer agreement, and discriminant validity. *Journal of Research in Personality, 43*(1), 84–90. DOI:10.1016/j.jrp.2008.10.002

Steinmetz, H., Isidor, R., & Baeuerle, N. (2012). Testing the circular structure of human values: A meta-analytical structural equation modelling approach. *Survey Research Methods, 6*(1), 61–75.

Stellar, J., Manzo, V., Kraus, M., & Keltner, D. (2011). Class and compassion: Socioeconomic factors predict responses to suffering. *Emotion, 12*(3), 449–459. DOI:10.1037/a0026508

Stephens, N., Markus, H., & Phillips, T. (2013). Social class culture cycles: How three gateway contexts shape selves and fuel inequality. *Psychology, 65*(1), 611–634. DOI:10.1146/annurev-psych-010213–115143

Strack, M., & Dobewall, H. (2012). The value structure in socioeconomically less developed European countries still remains an ellipse. *Europe's Journal of Psychology, 8*(4), 587–602. DOI:10.5964/ejop.v8i4.505

Strus, W., Cieciuch, J., & Rowiński, T. (2014a). Circumplex structure of personality traits measured with the IPIP-45AB5C questionnaire in Poland. *Personality and Individual Differences, 71*, 77–82.

Strus, W., Cieciuch, J., & Rowiński, T. (2014b). The circumplex of personality metatraits: A synthesizing model of personality based on the big five. *Review of General Psychology, 18*, 273–286.

Stumpf, H. (1993). The factor structure of the Personality Research Form: A cross-national evaluation. *Journal of Personality, 61*, 27–48.

Thalmayer, A., & Saucier, G. (2014). The Questionnaire Big Six in 26 nations: Developing cross-culturally applicable Big Six, Big Five and Big Two inventories. *European Journal of Personality, 28*(5), 482–496. DOI:10.1002/per.1969

Thornhill, R., & Fincher, C. L. (2014). *The parasite-stress theory of values and sociality: Infectious disease, history and human values worldwide.* New York: Springer.

Thornhill, R., Fincher, C., Murray, D., & Schaller, M. (2010). Zoonotic and non-zoonotic diseases in relation to human personality and societal values: Support for the parasite-stress model. *Evolutionary Psychology, 8*(2), 151–169. DOI:10.1177/147470491000800201

Tooby, J., & Cosmides, L. (1995). *The psychological foundations of culture.* New York: Oxford University Press.

Tracey, T., & Rounds, J. (1993). Evaluating Holland's and Gati's vocational-interest models: A structural meta-analysis. *Psychological Bulletin, 113*(2), 229. DOI:10.1037/0033–2909.113.2.229

Trivers, R. (1971). The evolution of reciprocal altruism. *Quarterly Review of Biology, 46*, 35–57.

Turchin, P. (2015). *Ultrasociety: How 10,000 years of war made humans the greatest cooperators on Earth.* Chaplin, CT: Beresta Books.

Turchin, P., Currie, T., Turner, E., & Gavrilets, S. (2013). War, space, and the evolution of Old World complex societies. *Proceedings of the National Academy of Sciences, 110*(41), 16384–16389. DOI:10.1073/pnas.1308825110

Turkheimer, E. (2000). Three laws of behavior genetics and what they mean. *Current Directions in Psychological Science, 9*, 160–164.

Turkheimer, E., Pettersson, E., & Horn, E. (2011). A phenotypic null hypothesis for the genetics of personality. *Annual Review of Psychology, 65*(1), 515–540. DOI:10.1146/annurev-psych-113011–143752

Uher, J. (2008a). Comparative personality research: Methodological approaches. *European Journal of Personality, 22*(5), 427–455. DOI:10.1002/per.680

Uher, J. (2008b). Three methodological core issues of comparative personality research. *European Journal of Personality*, 22, 475–496. DOI:10.1002/per.688

Uher, J, & Visalberghi, E. (2016). Observations versus assessments of personality: A five-method multi-species study reveals numerous biases in ratings and

methodological limitations of standardised assessments. *Journal of Research in Personality, 61*, 61–79.
Uzefovsky, F., Döring, A., & Knafo-Noam, A. (2016). Values in middle childhood: Social and genetic contributions. *Social Development*, 25(3), 482–502. DOI:10.1111/sode.12155
Valchev, V., van de Vijver, F., Rothmann, S., & Meiring, D. (2013). The use of traits and contextual information in free personality descriptions across ethnocultural groups in South Africa. *Journal of Personality and Social Psychology, 104*(6), 1077–1091. DOI:10.1037/a0032276
Van de Vijver, F., & Leung, K. (1997). *Methods and data analysis for comparative research*. Thousand Oaks: Sage.
Van de Vijver, F., & Poortinga, Y. H. (2002). Structural equivalence in multilevel research. *Journal of Cross-Cultural Psychology, 33*(2), 141–156. DOI:10.1177/0022022102033002002
Van de Vliert, E. (2007). Climatoeconomic roots of survival versus self-expression cultures. *Journal of Cross-Cultural Psychology, 38*(2), 156–172. DOI:10.1177/0022022106297298
Van de Vliert, E. (2009). *Climate, affluence, and culture*. New York: Cambridge University Press.
Van de Vliert, E. (2013a). Climato-economic habitats support patterns of human needs, stresses, and freedoms. *Behavioral and Brain Sciences, 36*(05), 465–480. DOI:10.1017/S0140525X12002828
Van de Vliert, E. (2013b). White, gray, and black domains of cultural adaptations to climato-economic conditions. *Behavioral and Brain Sciences, 36*, 503–521.
Van den Berg, S. M., de Moor, M. H., Verweij, K. J., Krueger, R. F., Luciano, M., Vasquez, A., . . . Boomsma, D. I. (2016). Meta-analysis of genome-wide association studies for Extraversion: Findings from the Genetics of Personality Consortium. *Behavior Genetics, 46*(2), 170–182. DOI:10.1007/s10519-015-9735-5
Vandenberg, R., & Lance, C. (2000). A review and synthesis of the measurement invariance literature: Suggestions, practices, and recommendations for organizational research. *Organizational Research Methods*, 3, 4–70.
Vauclair, C., & Fischer, R. (2011). Do cultural values predict individuals' moral attitudes? A cross-cultural multilevel approach. *European Journal of Social Psychology, 41*(5), 645–657. DOI:10.1002/ejsp.794
Vauclair, C., Hanke, K., Fischer, R., & Fontaine, J. (2011). The structure of human values at the culture level: A meta-analytical replication of Schwartz's Value Orientations using the Rokeach Value Survey. *Journal of Cross-Cultural Psychology, 42*(2), 186–205. DOI:10.1177/0022022110396864
Vecchione, M., Döring, A., Alessandri, G., Marsicano, G., & Bardi, A. (2016). Reciprocal relations across time between basic values and value-expressive behaviors: A longitudinal study among children. *Social Development*, 25(3), 528–547. DOI:10.1111/sode.12152
Vecchione, M., Schwartz, S., Alessandri, G., Döring, A., Castellani, V., & Caprara, M. (2016). Stability and change of basic personal values in early adulthood: An 8-year longitudinal study. *Journal of Research in Personality, 63*, 111–122. DOI:10.1016/j.jrp.2016.06.002

Verweij, K., Yang, J., Lahti, J., Veijola, J., Hintsanen, M., Pulkki-Råback, L., . . . Zietsch, B. (2012). Maintenance of genetic variation in human personality: Testing evolutionary models by estimating heritability due to common causal variants and investigating the effect of distant inbreeding. *Evolution*, *66*(10), 3238–3251. DOI:10.1111/j.1558–5646.2012.01679.x

Wacker, J., & Smillie, L. (2015). Trait Extraversion and dopamine function. *Social and Personality Psychology Compass*, *9*(6), 225–238. DOI:10.1111/spc3.12175

Wacker, J., Chavanon, M.-L., & Stemmler, G. (2006). Investigating the dopaminergic basis of extraversion in humans: A multilevel approach. *Journal of Personality and Social Psychology*, *91*(1), 171–187. DOI:10.1037/0022–3514.91.1.171

Walker, S., Wachs, T., Grantham-McGregor, S., Black, M., Nelson, C., Huffman, S., . . . Richter, L. (2011). Inequality in early childhood: Risk and protective factors for early child development. *The Lancet*, *378*(9799), 1325–1338. DOI:10.1016/s0140–6736(11)60555–2

Ward, T., & Gannon, T. (2006). Rehabilitation, etiology, and self-regulation: The comprehensive good lives model of treatment for sexual offenders. *Aggression and Violent Behavior*, *11*, 77–94.

Ward, T., Mann, R., & Gannon, T. (2007). The good lives model of offender rehabilitation: Clinical implications. *Aggression and Violent Behavior*, *12*(1), 87–107. DOI:10.1016/j.avb.2006.03.004

Welzel. (2010). How selfish are self-expression values? A civicness test. *Journal of Cross-Cultural Psychology*, *41*(2), 152–174. DOI:10.1177/0022022109354378

Welzel, C. (2013). *Freedom rising:Human empowerment and the quest for emancipation*. Cambridge: Cambridge University Press.

Welzel, C., & Inglehart, R. (2010). Agency, values, and well-being: A human development model. *Social Indicators Research*, *97*(1), 43–63. DOI:10.1007/s11205–009–9557-z

Welzel, C., Inglehart, R., & Klingemann, H. (2003). The theory of human development: A cross cultural analysis. *European Journal of Political Research*, *42*(3), 341–379. DOI:10.1111/1475–6765.00086

Wilkowski, B., & Ferguson, E. (2014). Just loving these people: Extraverts implicitly associate people with reward. *Journal of Research in Personality*, *53*, 93–102. DOI:10.1016/j.jrp.2014.08.006

Williams, L, Labonte, R, & O'Brien, M. (2003). Empowering social action through narratives of identity and culture. *Health Promotion International*, *18*, 33–40.

Wrzus, C., & Roberts, B. W. (2016). Processes of personality development in adulthood: The TESSERA framework. *Personality and Social Psychology Review*, *21*, 253–277.

Yang, Y., Read, S. J., Denson, T. F., Xu, Y., Zhang, J., & Pedersen, W.C. (2013). The key ingredients of personality traits: Situations, behaviors, and explanations. *Personality and Social Psychology Bulletin*, *40*(1), 79–91. DOI:10.1177/0146167213505871

Yarkoni, T. (2015). Neurobiological substrates of personality: A critical overview. In M. Mikulincer (Ed.), *APA handbook of personality and social psychology*, Vol 4, Personality processes and individual differences. Washington, DC: APA Press.

Zhu, B., Chen, C., Moyzis, R., Dong, Q., & Lin, C. (2015). Educational attainment-related loci identified by GWAS are associated with select personality traits and mathematics and language abilities. *Personality and Individual Differences*, *72*, 96–100.

Zuk, M. (1992). The role of parasites in sexual selection: Current evidence and future directions. *Advances in the Study of Behavior*, *21*, 39–68.

Zuk, M. (2013). *Paleofantasy: What evolution really tells us about sex, diet, and how we live*. New York: W. W. Norton.

INDEX